Abebe Shimeles
Audrey Verdier-Chouchane
Amadou Boly
Editors

Building a Resilient and Sustainable Agriculture in Sub-Saharan Africa

Editors
Abebe Shimeles
African Development Bank Group
Abidjan, Côte d'Ivoire

Amadou Boly
African Development Bank Group
Abidjan, Côte d'Ivoire

Audrey Verdier-Chouchane
African Development Bank
Abidjan, Côte d'Ivoire

https://doi.org/10.1007/978-3-319-76222-7

Foreword

Agriculture is key to Africa's prosperity and development. The sector now employs more than half of the labor force but produced only 15% of GDP continent-wide, with considerable cross-country variation, indicating that current performance remains below potential. *Building a resilient and sustainable agriculture in Sub-Saharan Africa* contains technically rich and policy-relevant chapters that shed light on the performance and challenges of Africa's agriculture sector, particularly, climate shocks exposure due to an over-reliance on rain-fed agriculture combined with low investment levels.

This book offers an excellent contribution for better understanding of the current state and prospects of Africa's agricultural transformation and generates valuable insights for improved agricultural policymaking on the continent. Beyond the collection of and reflection on African stories, this compilation interestingly proposes pragmatic policy options for improving agriculture productivity, mitigating climate shocks exposure and promoting agro-industrialization. As advocated by the authors, these three targets are key to unleashing Africa's agriculture potential.

It is now time for Africa to feed Africa … and the rest of the world. And achieving this goal will require synergetic actions between governments, private sector or other stakeholders. The current collection of analytical chapters from African researchers is an excellent contribution to that collective effort.

Cornell University
Ithaca, NY, USA

Christopher B. Barrett,
December 2017

Acknowledgments

This book brings together a selection of chapters presented at the 2016 African Economic Conference (AEC) in Abuja, Nigeria, under the theme: "Feed Africa: Towards Agro-Allied Industrialization for Inclusive Growth". The chapters have been selected by experts in agriculture at Cornell University and then enriched by valuable comments by peer reviewers, discussants and rapporteurs at the conference. Our appreciation goes to Charlotte Karagueuzian for excellent research assistance and coordination with the authors and to the African Development Bank Language Services Department for translating some of the chapters from French to English.

The AEC 2016 provided a unique opportunity for policymakers and researchers, including the diaspora, to interact and debate the critical policy issues affecting Africa. This conference would not have been possible without the full support and the leadership of the African Development Bank's President, the Executive Secretary of the Economic Commission for Africa and the Administrator of the United Nations Development Programme. Many staff members of these institutions dedicated their time and energy in making the conference a success. Their contributions are deeply appreciated.

We are extremely grateful to the researchers who presented their papers at the AEC 2016 for their high-quality contribution. This publication aims at stimulating debate and shaping policy recommendations around the resilience and sustainability of agriculture in sub-Saharan Africa.

Contents

Notes on Contributors

Tahirou Abdoulaye is an agricultural economist at the International Institute for Tropical Agriculture (IITA), Ibadan, Nigeria. He has been involved in evaluation and impact assessment of several projects, mainly in West Africa. He holds a PhD degree in Agricultural Economics. He has produced papers on impact assessment of research activities in several African countries including Niger, Ghana, Mali, Senegal, Nigeria and Benin. His research work covers a wide range of rural economic issues including seed systems, farm-level efficiency and also technology evaluation and transfer. His more recent research interest focuses on innovation systems and how they can help increase technology uptake by small farmers.

D. P. K. Amegashie is a graduate from the University of Ghana, Legon, with a Master of Philosophy degree in Agricultural Economics in 1995. He is a senior lecturer in the Department of Agricultural Economics and Agribusiness and teaches Introduction to Economics, Farm Management, Project Analysis, Research Methods and Management Accounting. He has supervised several undergraduate dissertations and graduate theses in diverse areas of study. He served on various boards and committees in the University and is on a post-retirement contract in the Department.

Opeyemi E. Ayinde is an associate professor in the University of Ilorin, Nigeria, and senior research associate fellow in the South African Research Chair Initiative (SARCHI) Tshwane University of Technology, South Africa. She holds a PhD degree in Agricultural Economics. She is carrying out research on economics of innovation in maize varieties developed for drought-prone ecologies in Nigeria.

The research is awarded by IITA under Prime Agreement from The International Maize and Wheat Improvement Center (CIMMYT) for collaboration in Implementation of Stress Tolerant Maize for Africa (STMA)-Phase II. She has publications in peer-reviewed journals as well as presentation of academic papers at international conferences.

Amadou Boly is Principal Research Economist at the African Development Bank. His research interests are in development and experimental economics, and his articles have been published in leading international journals. Previously, Boly worked for United Nations University-World Institute for Development Economics Research (UNU-WIDER) (2014–2015, Finland) and United Nations Industrial Development Organization (UNIDO) (2009–2014, Austria), mainly in technical assistancc and global forum activities related to industrial statistics and industrial policy. He holds a PhD in economics from the University of Montreal, an MSc in International Business from the University of Groningen and a BSc in Economics and Management from the University of Ouagadougou.

Irene Susana Egyir is a senior lecturer in the University of Ghana at the Department of Agricultural Economics and Agribusiness. Her research is concerned with productivity and profitability of organic, conventional and sustainable farming systems. Her consultancies have focused on youth, gender and microfinance in agricultural value chains. Dr. Egyir has made a number of contributions in published articles, books, technical reports, discussion papers and conference proceedings across the continents. She has founded a development NGO named Multi-Features and Capacity Enhancing Services which links urban micro-entrepreneurs to Research, Extension, Finance, Consulting, Cooperative and Media (REFCCOM) Services.

Abdul-Fahd Fofana holds a PhD in International Economics. He conducted several research works on topics pertaining to global value chains. He worked as a research assistant at the Center for Research and Training in Economics and Management in the University of Lomé, Togo. He was also member of the research team in the project of the Poverty and Economic Policy (PEP) network supported by the Community-Based Monitoring System (CBMS) on the theme: "the willingness to pay by Togolese informal sector workers to access social protection". He was also responsible for the tutorials of the courses of microeconomics, macroeconomics and international trade at the Faculty of Economics and Management of the University of Lomé.

Namalguebzanga Christian Kafando has worked since 2013 as an international consultant for Trade Capacity Building and Quality projects at the UNIDO. He is providing technical assistance for development and implementation of projects related to International Standards, Capacity Building and Quality infrastructures. Kafando holds a Doctorate in Economics (highest honours, Auvergne University) with a specialisation on industrialisation, structural change and economic development of Africa and a Master's Degree in Macroeconomics and International Finance (honours, Nice Sophia Antipolis University).

His research interests include structural changes, industrialisation, extractive industries, exchange rate and exchange rate regime.

Francis H. Kemeze works as an economic consultant at the International Food Policy Research Institute. He has conducted research broadly, involving applied microeconometrics, behavioral economics, agricultural finance, water management, food security, trade and poverty, policy evaluation methods and value chain development. He has strong quantitative and econometric skills that can be applied to a wide spectrum of economic problems. He holds a BSc in Economics from Université de Dschang, an MPhil in Economics from Université de Yaounde II and a PhD in Applied Agricultural Economics and Policy from the University of Ghana. He is an expert on impact evaluation and food security measurement.

Lauretta Sandra Kemeze is a PhD candidate in Applied Agricultural Economics and Policy at the University of Ghana, Legon. Her PhD research describes how Jatropha adoption influences the level of crop incomes of farmers and their preferences for production contracts in Northern Ghana. She holds a Master's degree in Economics from the University of Yaounde II, Soa. Her research interests are rural development, bioenergy economics, choice modeling, technology adoption and food security. She has participated in several international conferences.

Boris Odilon Kounagbè Lokonon holds a PhD in Climate Change Economics under the West African Science Service Center on Climate Change and Adapted Land Use (WASCAL) program, from the Université Cheikh Anta Diop, Dakar, Senegal. He is a lecturer and researcher at the Faculté des Sciences Economiques et de Gestion, Université de Parakou, Benin, where he teaches Microeconomics, Agricultural Production Systems and Institutions, Environmental and Resource Economics and Applied Mathematics. His research interests are climate change economics, environmental and resource economics and agricultural policy.

Anani Nourredine Mensah holds a PhD in International Economics from the University of Grenoble 2, France. He also holds a PhD in Economics from the University of Rennes 1, France. He worked as the Deputy Dean of the Economics and Management Faculty at the University of Lomé, Togo, where he teaches International Trade and Microeconomics at both the graduate and undergraduate levels. He has conducted several research works on global value chains, cotton price change and welfare, debt and investment efficiency with regard to Islamic finance framework, natural resources and international trade, and monetary policy in West African Economic and Monetary Union (WAEMU) countries.

Akwasi Mensah-Bonsu is a senior lecturer at the Department of Agricultural Economics and Agribusiness, University of Ghana, Legon. He holds a PhD in Development Economics from Vrije Universiteit, Amsterdam, awarded in 2003. His areas of research include development economics and policy analysis for agriculture, modeling of the agricultural sector resources use and production efficiency analysis, cost-benefit analysis, project managing (including monitoring and evaluation), agricultural markets and environmental issues. He has undertaken a number of research as well as other extension works for both local and international agencies, including Ghana's Ministry of Food and Agriculture, FAO, DANIDA, UNDP, NEPAD and International Centre of Insect Physiology and Ecology (icipe).

Jean Hugues Nlom is a senior lecturer and Vice-Dean in charge of Programming and Follow-Up of Academic Affairs at the Faculty of Economics and Management Sciences (FEMS) of the University of Maroua (UMa), Cameroon. Dr. Nlom teaches Microeconomics, Environmental Economics and Mathematical Tools for Environmental Modeling. Before his appointment at the UMa, he was a senior lecturer at the FEMS of the University of Yaoundé II, Cameroon. Jointly with its academic career, he works as a consultant in international organizations in Cameroon and in Africa. He is also a member of the African Economic Research Consortium (AERC) network.

Gbadebo A. Olaoye is Professor of Plant Breeding at the University of Ilorin, Nigeria. He has supervised over 70 undergraduate and graduate students' degree projects. He is a recipient of many local and international awards including the Adult Higher Education Alliance (AHEA), USA; International Professional Award, Bill & Melinda Gates Foundation; and the Howard G. Buffett Foundation awards on Breeding and Promotion of Drought Tolerant Maize for Africa (DTMA) project. Professor Olaoye has published over 50 articles in local and international journals. He is the team leader of STMA project in southern guinea savanna, Nigeria.

Adeola O. Oloyede holds a PhD degree in Agricultural Economics. She is a research assistant in the team on STMA at the University of Ilorin in Nigeria, a research sponsored by the IITA. She has presented some of her works at international conferences and published her findings in local and international peer-reviewed journals. She has also received some grants in the form of travel support to attend international conferences. Adeola is a multidisciplinary researcher in the making. She is working on ways to promote environmental sustainability and issues relating to small-scale farmers.

Idrissa Ouiminga is a young teacher and researcher in Economics. He studied at University Ouaga 2, Burkina Faso and Laval University, Canada. He has extensively worked on environmental issues related to agriculture such as climate change and the preservation of natural resources.

Carren Pindiriri is a lecturer and a researcher at the University of Zimbabwe. In the lecturing position, he teaches Econometrics, Agricultural Production, Microeconomics, International Trade, Environmental, Growth and Development Economics. Before joining the University of Zimbabwe, Pindiriri worked for the Infrastructure Development Bank of Zimbabwe as a business development officer. He holds BSc and MSc in Economics from the University of Zimbabwe and an MPhil in Environmental and Development Economics from Oslo University. His research interests include agricultural growth and technological adoption by farmers in Africa, environmental economics, and industrialization of African economies.

Adedoyin Mistura Rufai has a PhD degree in Agricultural Economics from the University of Ibadan. She teaches part time at the Federal College of Animal Health and Production Technology, Ibadan, Nigeria, where she has taught modules in research methods, farm management and mathematics in agriculture. She is actively involved in research and her areas of specialization include gender and welfare analysis, health economics, agricultural production and profitability analysis, impact assessments, rural and developmental economics. She is a member of the Nigerian Association of Agricultural Economists (NAAE).

Mutiat Bukola Salawu is a PhD student in the Department of Agricultural Economics, University of Ibadan, Nigeria. She is also a lecturer at the Department of Agricultural Extension and Management of Federal College of Animal Health and Production Technology, Ibadan, Nigeria. She has taught courses such as Farm Management, Macroeconomics, Farm Business Organization, Agricultural Marketing, Agricultural Finance, Microeconomics, Agricultural Policy, Agricultural Mathematics and Research Methodology. Her research interests are gender analysis, health and developmental economics. She is a member of Agricultural Society of Nigeria, Nigerian Institute of Management and NAAE.

Kabir Kayode Salman holds a PhD degree in Agricultural Economics from the University of Ibadan, Nigeria. He is a senior lecturer in the same department. He has taught Resource Economics, Human Capital Management, Econometrics and Research Methodology. His research interests are rural economics, political economy, impact assessment, and policy and development. He had worked with International Food Policy Research Institute (IFPRI), AERC and Forum for Agricultural Research in Africa (FARA). He is a member of AGRODEP and NAAE.

Abebe Shimeles is Acting Director of the Macroeconomic Policy, Forecasting and Research Department at the African Development Bank. Previously Shimeles has worked for The World Bank, UNECA, ACTIONAID and Addis Ababa University, Ethiopia, in different capacities. His recent research interest includes labor market integration, migration issues in Africa and impact evaluation of policy interventions. He holds a PhD in Economics from the University of Goteborg, MSC from Delhi School of Economics and undergraduate degree in economics from Addis Ababa University.

Christelle Tchamou Meughoyi holds a Master's in Economics, with a focus on Economics Mathematics and Econometrics and is a PhD student at the University of Yaounde II in Cameroon. An expert in quantitative techniques and issues related to agriculture, she has written two working papers on the "Typology of maize-based family farms in the Central and Western regions of Cameroon" and the "Characterization of maize-producing households in the agro-ecological regions of Central and West Cameroon". She is a member of the Research Laboratory in Applied Microeconomics (REMA) whose Director is the Rector of the University of Dschang in Cameroon and her PhD supervisor. In addition, she also carries out work supervision and teaching activities in financial mathematics, statistics and general economics.

Joseph Tinarwo is the coordinator for the Politics and Public Administration Programme in the Julius Nyerere School of Social Sciences at Great Zimbabwe University in Zimbabwe. His research interests include agriculture, food and nutrition security, public policy and governance. Over the years, Tinarwo has been actively involved in various programs as a development consultant, with key result areas including institutional capacity assessments, training, monitoring and evaluation, designing and managing special studies.

Audrey Verdier-Chouchane is Chief Research Economist at the African Development Bank. Prior to joining the African Development Bank, Verdier-Chouchane taught Macroeconomics and Development Economics at the University of Nice Sophia Antipolis, France, from where she holds a PhD in Economics. She has mainly published on the analysis and measurement of poverty and inequalities, with a special focus on gender disparity and labor market issues for the youth.

List of Figures

List of Graphs

List of Tables

1

Introduction: Understanding the Challenges of the Agricultural Sector in Sub-Saharan Africa

Abebe Shimeles, Audrey Verdier-Chouchane, and Amadou Boly

1.1 Introduction

Sub-Saharan African countries have recorded relatively high economic growth over the past two decades, but this growth has largely been jobless and poverty and inequality are still widespread (African Development Bank et al. 2017). Also, Africa's growth has hardly been accompanied by structural transformation. The labor force is still stuck in subsistence production and low productive agricultural sector which employs more than half of the sub-Saharan population. According to the International Labour Organization (2017), the agricultural sector employs an average of 54% of the working population in Africa. In Burundi, Burkina Faso and Madagascar, more than 80% of the labor force works in agriculture. By contrast, in Angola, South Africa, and Mauritius, the agricultural sector only employs 5.1%, 4.6%, and 7.8% of the population, respectively.

A. Shimeles • A. Verdier-Chouchane (✉) • A. Boly
African Development Bank, Abidjan, Côte d'Ivoire

A. Shimeles et al. (eds.), *Building a Resilient and Sustainable Agriculture in Sub-Saharan Africa*, https://doi.org/10.1007/978-3-319-76222-7_1

In addition to the massive labor force, sub-Saharan Africa has also the highest area of arable uncultivated land in the world and huge agricultural growth potential (Kanu et al. 2014). But countries have not yet taken advantage of it. Despite the importance of the sector, about one-fourth of the population experiences hunger in sub-Saharan Africa. Out of about 795 million people suffering from chronic undernourishment globally, 220 million live in sub-Saharan Africa. At around 23.2%, the Food and Agricultural Organization (FAO) (2015) indicates that this is the highest prevalence of undernourishment worldwide. Even in abundant regions, food shortages can happen, mostly due to poor conservation techniques or post-harvest losses. In fact, the continent overall is a net importer of food which puts additional strain on scarce foreign exchange reserves.

In agriculture, women face particularly severe challenges. Although they represent 47% of the labor force, they are prominently smallholder farmers because the patriarchy system has tended to discriminate against them (Kanu et al. 2014). Customary laws and rules governing ownership and transfer of land rights are generally unfavorable to women in sub-Saharan Africa, conferring title and inheritance rights upon male family members. For Woldemichael et al. (2017), women in agriculture also experience lack of access to finance, modern inputs as well as lack of knowledge and skills of modern agricultural practices. Without these disadvantages as compared to males, women could be as industrious as men, not only in agriculture but also in every sector. As per the FAO (2011), if women had access to the same resources as men, their agricultural yields would increase by up to 30%, reducing by 100–150 million the number of hungry people globally.

For Moyo et al. (2015), Africa's low use of irrigation and overwhelming dependence on rain-fed agriculture explain the continent's low agricultural productivity. The main staples of sub-Saharan Africa are unirrigated crops (maize, cassava, millets, sorghum, yams, sweet potatoes, plantains and rice). In addition, limited public funding in the agricultural sector has also prevented the provision of adequate institutional support and suitable business environment, in turn hindering private sector participation and investment in agriculture. Under the terms of the New Partnership for Africa's Development's (NEPAD) Comprehensive Africa Agriculture Development Programme (CAADP),[1] governments should devote 10% of national spending to agriculture in order to support water management, intensify irrigation, reduce the continent's dependence on rain-fed agriculture and increase resilience to climate change. However, Fig. 1.1 clearly indicates that public expenditure in agriculture is far below the 10% target, ranging from a low 0.15% in Guinea-Bissau to 3.61% in Malawi.

This low level of investment prevents countries from adapting to climate change shocks, limited rainfall and weather shocks. For Kanu et al. (2014), climate change poses a great challenge to promoting inclusive growth in sub-Saharan Africa. Due to climate disturbances, growing and harvesting seasons are increasingly unpredictable. In addition, the region has one of the most vulnerable agriculture worldwide to extreme weather events, such as drought and floods due to its heavy reliance on rain-fed agriculture, low adaptive capacity and limited infrastructure development.

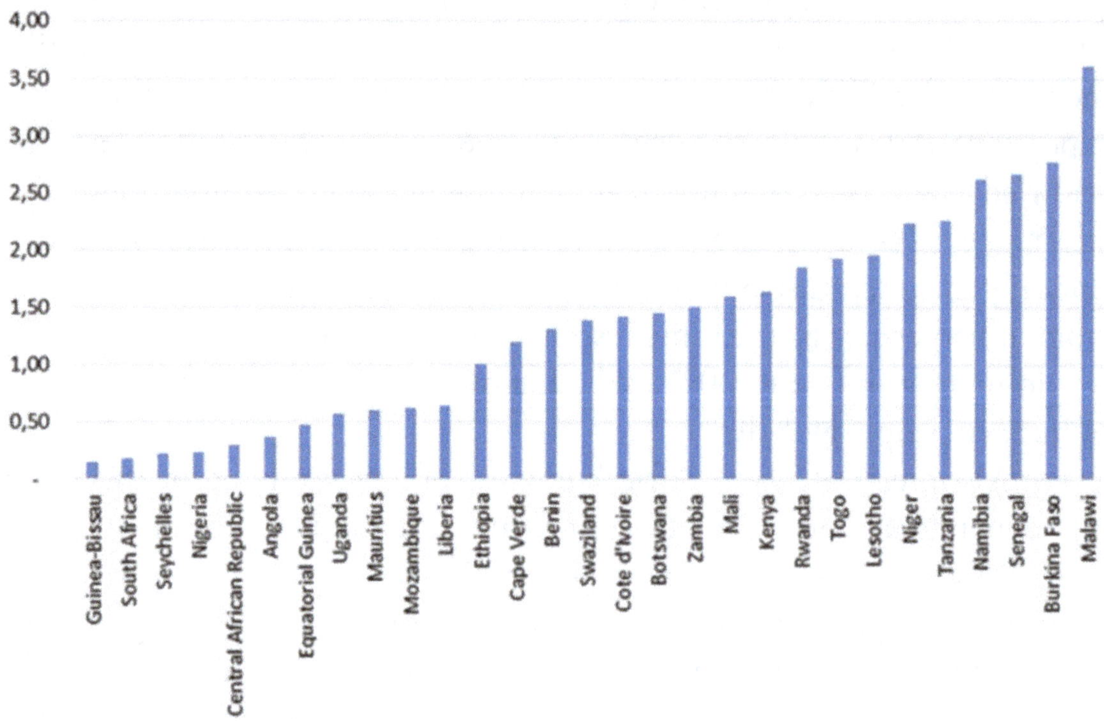

Fig. 1.1 Percentage of agricultural expenditure in total GDP. Source: Authors, based on IFPRI (2017). Note: more recent year between 2009 and 2012

Increasing agricultural productivity would support structural transformation process and economic growth in sub-Saharan Africa for three reasons. First, it will enable the labor force to move from the agricultural sector into other sectors and help developing the manufacturing and services sector. There is a high momentum behind developing and promoting the agricultural sector as a catalyst to industrialization and agribusiness development. Second, it will allow African farmers to better manage and integrate the entire agricultural value chain from the farm to storage, transport, processing, marketing and distribution. This will not only improve food supply but also create additional revenues and jobs. Third, farmers will be able take advantage of large markets, increase trade and exports of agricultural products and progressively integrate regional and global value chains (GVCs).[2] However, farmers need to deliver high-quality products at competitive prices and integrate international distribution channels by satisfying the norms and standards set out by their trading partners (AfDB 2014).

This volume on *Building a Resilient and Sustainable Agriculture in Sub-Saharan Africa* presents 13 chapters related to a better understanding of agriculture in sub-Saharan Africa and to the best policy options for enhancing its resilience and sustainability. In Part I, the authors looked into the issues of productivity, sometimes by gender, with a specific focus on modern inputs, including machinery, fertilizers and improved seed varieties. They also demonstrate that some crop adoption can reduce farmers' income if not adequately planned or completed with other measures. In Part II, the authors analyzed the climate change challenges in agriculture and its vulnerability to drought and declining soil fertility. Authors deal with soil and water conservation techniques, land tenure issues and weather index insurance. In Part III, authors considered the promotion of agro-industrialization. They review the contribution of agricultural activities to the development of the manufacturing sector, focus on the creation of special economic zones (SEZs) to transform agriculture and examine Economic Community of West African States' (ECOWAS) integration in GVCs. The remaining of this overview chapter gives a flavor of the main issues discussed. It follows the outline of the entire volume.

1.2 Improving Agricultural Productivity

In sub-Saharan Africa, agricultural sector productivity considerably lags other regions (see Fig. 1.2). According to NEPAD (2013), Africa has 33 million farms of less than 2 hectares, accounting for 80% of all farms. Given that the farming system mainly relies on family's capital and labor force for production, the overall productivity is low. Subsistence farmers cannot significantly contribute to food security at a national scale because the pieces of land they have access to are too small (Africa Research Institute 2009).

In Chap. 2, **Christelle Tchamou Meughoyi** analyzes whether improved maize seeds can significantly increase the productivity of family farms in Cameroon. Considering 259 family farms, she uses the Blinder-Oaxaca decomposition technique to estimate the difference in productivity between the adopters and non-adopters of improved seeds. She concludes that yield obtained by adopters is 1.42 times higher than

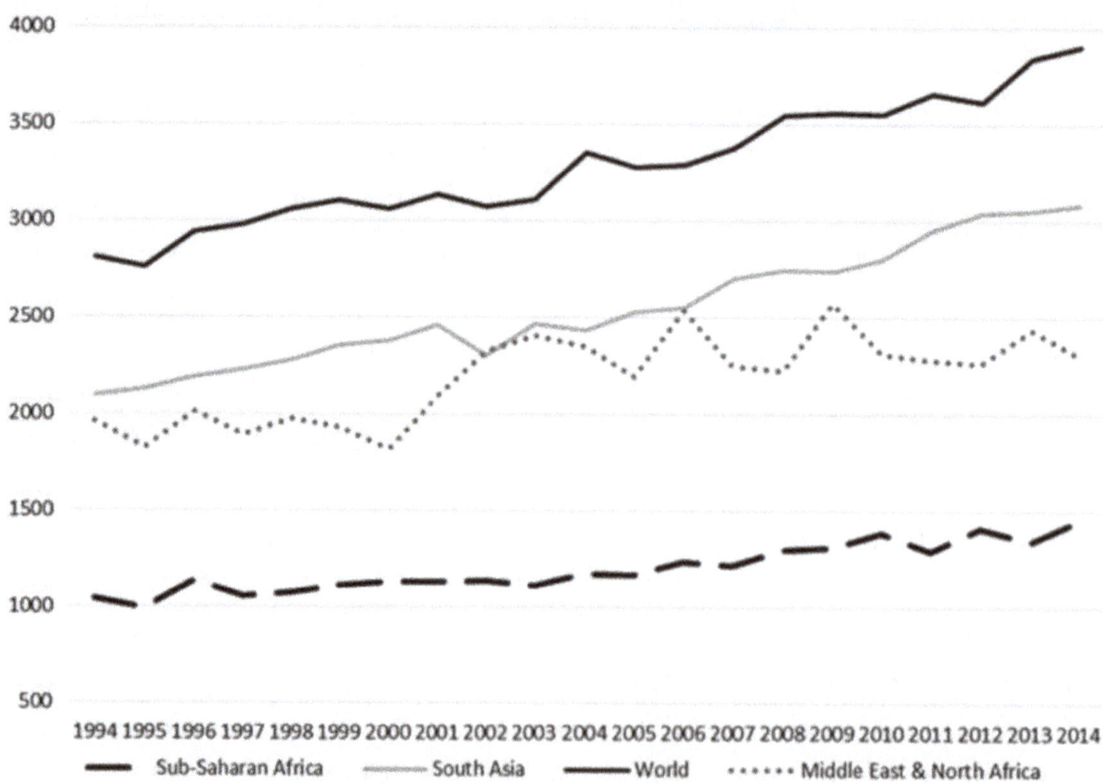

Fig. 1.2 Cereal yield (kilogram per hectare). Source: Authors based on data from the World Bank (online) *World Development Indicators*

the one obtained by the non-adopters. However, this is far below the expected theoretical productivity gap. Farmers may fail to comply with the conditions associated with improved seed varieties, including the use, method and period of application of fertilizers, herbicides and weeding. She highlights the fact that such innovations should be combined with other modern inputs to have the expected result. Ripple effect of improved seed varieties on other factors of production entail additional cost for the farmers and should be taken into account.

However, the modernization of agriculture is a priority for increasing productivity and moving away from subsistence agriculture. In Chap. 3, **Carren Pindiriri** examines the drivers of agricultural technology adoption by smallholder farmers in Zimbabwe. Using a sample of 411 farmers in Hurungwe, she finds a population technological gap of 12.7% resulting from lack of awareness. Farmers' propensity to adopt modern technologies increases with education, training, access to credit and income. She then highlights the need to reduce technology information asymmetry among farmers through various media and to improve financial services in rural areas.

In Chap. 4, **Adedoyin Mistura Rufai, Kabir Kayode Salman and Mutiat Bukola Salawu** explore the influences of input utilization on labor productivity among men and women in Nigeria using the General Household Survey and a quantile regression method. The use of modern inputs, such as fertilizer, herbicide, pesticide, animal traction and machinery equipment, is generally low in Nigeria's agriculture. However, women, who constitute about half of the labor force in the agricultural sector, face lower access than men. As a result, it contributes to gender productivity differentials and ineffective use of the vast human resources available. The authors conclude that productivity of the agricultural sector in Nigeria can be improved extensively through gender-sensitive policy and capacity building of female farmers.

In Chap. 5, **O. E. Ayinde, T. Abdoulaye, G. A. Olaoye and A. O. Oloyede** also consider the role of women in Nigeria's agriculture with the objective of increasing their involvement in on-farm trails of agricultural technologies. Authors interviewed about 80 female farmers who have adopted improved seeds (drought-tolerant (DT) maize varieties).

The women's varietal preference and the profitability of the maize varieties differ across locations. However, the women farmers ranked the DT maize variety as the best, having the highest profit and returns to investment at all the locations. The authors reaffirm the specific role of women in ensuring food security and increasing agricultural productivity. They recommend their strong involvement in the development and testing of agricultural innovation, especially in a context of climate change.

In Chap. 6, **Lauretta S. Kemeze, Akwasi Mensah-Bonsu, Irene S. Egyir, D. P. K. Amegashie and Jean Hugues Nlom** focus on the impact of the Jatropha curcas adoption on farmers' incomes in Northern Ghana. In the mid-2000s, this bioenergy cultivation was presented as a panacea and a promising feedstock for biofuels. However, large-scale Jatropha (100 hectares and more) development was criticized for land grabbing and food insecurity. In addition, using data from 400 farmers and a Propensity Score Matching method, the authors found that the adoption of Jatropha curcas significantly reduces farmers' total crop incomes per hectare. The gender analysis also reveals that this reduction was more important for female-headed households. The authors recommend the protection of rural people and food crops and insist in the necessity of adopting a proper regulation of the biofuel sector in Ghana.

1.3 Addressing Climate Change Challenges

A key challenge for sub-Saharan African countries is to reverse costly environment degradation and better adapt to climate change shocks. AfDB (2011) estimated that an investment of USD 20–30 billion annually would be required to reduce climate vulnerability in Africa and to maintain the potential negative effect at 1.8% of Africa's GDP. However, sometimes there is a trade-off between protecting the environment and enhancing agricultural productivity (Verdier-Chouchane & Karagueuzian 2016).

In Chap. 7, **Idrissa Ouiminga** proposes solutions to combat land degradation and desertification. Repeated droughts and inadequate practices in agriculture have resulted in the decline of soil fertility and the degradation of the vegetation cover. Considering different soil and water conservation techniques, the author examines both financial profitability, yields and social impact (labor) through a cost-benefit analysis in the municipality of Yalgo, Burkina Faso. Soil conservation techniques range from mulching, compost, manure and fertilizer such as NPK (Nitrogen, Phosphorous and Potassium), while water conservation techniques include dugs and constructed structures such as stony ropes, half-moons and water cuvettes. These techniques, usually supported by public investment, represent a good alternative to adapting to climate change.

In Chap. 8, **Boris Odilon Kounagbè Lokonon** analyzes the vulnerability of villages to climate shocks and the extent to which land tenure has affected vulnerability in Benin. Using data over 1998–2012, the author calculates indices for each dimension of vulnerability (adaptive capacity, sensitivity and exposure) and then overall vulnerability indices. Generally, the situation has improved over the period but the adaptive capacity is very low, questioning the villages' resilience on future climate shocks. The econometric analysis reveals that farmers' labor sharing and organizations have the potential to lessen vulnerability to climate shocks. In contrast, land tenure is not significant in strengthening resilience. It is considered as a social protection measure that could increase productivity should it be accompanied by appropriate financial capital and access to technology.

In Chap. 9, **Francis H. Kemeze** looks at the weather index insurance which partially protects farmers against climate variability and partially compensates for the negative effects of drought. Specifically, the author looks into the effects of drought index insurance on the demand for supplemental irrigation in Northern Ghana. Weather insurance does not usually cover actual on-farm losses and does not replace the crop loss. In addition, in case of drought, price of staple food goes up and prevents farmers from smoothing their consumption with the insurance premium. The result of the randomized control trials (RCT) analysis confirms that index insurance covers the costs of irrigations in drought years. As a result, farmers should apply additional water to otherwise rain-fed crops to save the harvest. The investment in water management is complementary drought adaptation strategy.

1.4 Promoting Agro-Industrialization

The connection of small-scale farmers to large business farmers through mutually beneficial contract farming (also called out-grower schemes) facilitates famers' access to inputs, financing, end-markets as well as their participation in agriculture value chains. However, for Verdier-Chouchane & Karagueuzian 2016, a comprehensive transformation of the agricultural sector in Africa toward agro-industrialization requires investments in technology and innovation in order to improve the productivity of both land and labor. An example of the use of Internet to improve integration into value chains is the traceability of food and animals which entails displaying the lot number and the production facility name on each case of the product and recording this information on invoices and bills of lading (Verdier-Chouchane & Boly 2017). Innovation can also facilitate commercialization and enhance farmers' access to broader markets and financial systems. For Moyo et al. (2015), this will ultimately allow the creation of modern integrated agribusiness value chain economies based on specialization. Domestic and international private investment in the agricultural sector has already caused remarkable changes in agribusiness in Africa with positive effects on smallholder farmers' revenues and productivity (Kanu et al. 2014).

In Chap. 10, **Namalguebzanga C. Kafando** considers Africa's z advantage in agricultural products and its industrialization based on the exploitation of natural resources. The author confirms that their processing can enhance the value added of exports and Africa's industrial development, especially in West and Central Africa where the value added of the manufacturing sector is very low. However, the author also reviews the obstacles toward industrial and regional value chains development in Africa and recommends some policy actions. He mainly focuses on the role of education and skills, transport infrastructure, governance quality, trade integration and the use of technology.

In Chap. 11, **Joseph Tinarwo** focuses on the role that SEZs can play in transforming Africa's agriculture and developing agribusiness markets. In SEZs, economic regulations are different from those of the rest of the country. By mainstreaming the administration, providing tax incentives and low tariffs, SEZs improve the business environment and facilitate the access to new markets and encourage the concentration of industrial growth. Basing his analysis on experiences in Asia and Africa, the author provides recommendations to ensure a successful agricultural transformation in Africa through SEZ development. He also highlights the need for further research and additional empirical data.

Finally, in Chap. 12, **Anani N. Mensah and Abdul-Fahd Fofana** examine the level of integration and the upgrading of ECOWAS member countries in the GVCs. Using indicators of product sophistication and diversification, the authors find that the agricultural products ECOWAS countries export abroad are mainly primary low-tech products. Trade is strongly driven by forward integration as primary products are transformed and used for manufacturing outside ECOWAS. In contrast, countries should develop backward integration by improving competitiveness`, supporting export companies and investing in infrastructure.

Notes

1. http://www.nepad.org/cop/comprehensive-africa-agriculture-development-programme-caadp.
2. For Gereffi and Fernandez-Stark (2011), the global value chain is the interconnected production process that goods and services undergo from conception and design through production, marketing and distribution. Country's participation in GVC trade is measured by both the backward and forward integration. Backward integration occurs when a country sources foreign inputs for its export production, while forward integration occurs when a country provides inputs for a foreign country's export production.

References

AfDB. (2014). *Global Value Chains and Africa's Integration into the Global Economy*. Annual Report 2013, Chapter 2. Tunis: AfDB.

AfDB, Organisation for Economic Co-operation and Development [OECD], & United Nations Development Programme [UNDP]. (2017). *African Economic Outlook 2017*. Paris: OECD Publishing. Retrieved from http://www.africaneconomicoutlook.org/en/.

Africa Research Institute. (2009, November). *Waiting for a Green Revolution*. Briefing Note 0902.

African Development Bank [AfDB]. (2011). *The Cost of Adaptation to Climate Change in Africa*. Tunis: African Development Bank.

FAO. (2015). *The State of Food Insecurity in the World*. Rome: FAO.

Food and Agricultural Organization of the United Nations [FAO]. (2011). *The State of Food and Agriculture 2010–2011: Women in Agriculture. Closing the Gender Gap for Development*. Rome: FAO.

Gereffi, G., & Fernandez-Stark, K. (2011). *Global Value Chain Analysis: A Primer*. Durham, NC: Center of Globalization, Governance and Competitiveness.

IFPRI. (2017). *SPEED Dataset 2015*. Washington: IFPRI.

International Labor Organization [ILO]. (2017). *Foresight Africa 2017 Report*. Geneva: ILO.

Kanu, B. S., Salami, A. O., & Numasawa, K. (2014). *Inclusive Growth: An Imperative for African Agriculture*. Tunis: African Development Bank.

Moyo, J. M., Bah, E. M., & Verdier-Chouchane, A. (2015). Transforming Africa's Agriculture to Improve Competitiveness. *World Economic Forum, World Bank and AfDB (2015)*. Africa Competitiveness Report 2015. Geneva: WEF.

NEPAD. (2013). *Agriculture and Africa—Transformation and Outlook*. Johannesburg: NEPAD.

Verdier-Chouchane, A., & Boly, A. (2017). Introduction: Challenges to Africa's Agricultural Transformation. *African Development Review, 29*(S2), 75–77.

Verdier-Chouchane, A., & Karagueuzian, C. (2016). Moving Towards a Green Productive Agriculture in Africa: The Role of ICTs. *Africa Economic Brief, 7*(7). Côte d'Ivoire: African Development Bank.

Woldemichael, A., Salami, A., Mukasa, A., Simpasa, A., & Shimeles, A. (2017). Transforming Africa's Agriculture Through Agro-Industrialization. *Africa Economic Brief, 8*(7), Abidjan: African Development Bank.

Part I

Improving Agricultural Productivity

2

Improved Seeds and Agricultural Productivity of Family Farms in Cameroon

Christelle Tchamou Meughoyi

2.1 Introduction

Like in many Sub-Saharan African countries, Cameroon has an economy dominated by agriculture which employs nearly 60% of the labour force, generates about 15% of budgetary resources and accounts for 30% of GDP (World Bank 2008). In addition, the agricultural sector has the highest ripple effect on the other sectors, thereby contributing significantly to poverty reduction.

Agriculture is still dominated by family farms,[1] which are mainly located in rural areas (ECAM 3 2007). Family farms play a vital role in agricultural development. In Cameroon, they supply nearly 95% of the food products (cocoyam, sweet potato, maize, etc.) and retain about 80% of their production for on-farm consumption (ACDIC 2008). Despite their importance, family farms always face numerous challenges, including low productivity (World Bank 2008; Mugisha and Diiro 2010).

C. Tchamou Meughoyi (✉)
Mathematical Economics and Econometrics,
Université de Yaoundé II, Soa, Cameroon

A. Shimeles et al. (eds.), *Building a Resilient and Sustainable Agriculture in Sub-Saharan Africa*, https://doi.org/10.1007/978-3-319-76222-7_2

The low productivity of family farms is due to the numerous problems facing the agricultural sector. These include the poor organization of actors, biotic constraints (diseases and pests), road infrastructure-related constraints (degraded rural roads and obsolescent and inadequate means of transportation) and production-related constraints (limited and high cost of good quality seeds, small crop areages (or areas), etc.).

Many solutions have been proposed to boost agricultural productivity, in particular improvement of infrastructure, supply of fertilizers, implementation of institutional reforms, as well as the introduction of innovations (Mead 2003; Fan et al. 2004; Scarpetta and Tressel 2004; Pedelahore and Tchatchoua 2010; Awotide et al. 2012; Adofu et al. 2013). Innovations help to improve agricultural performance, provide benefits and bring about social change. Furthermore, many studies have confirmed that innovations can serve as a powerful lever for improving the productivity of family farms (Edwin and Master 2005; Ntsama 2007; Ogunniyi and Kehinde 2015).

However, it should be noted that the concept of innovation comprises various aspects; hence, one can talk of institutional, political, social, organizational innovations, as well as knowledge and practice innovations and material innovations. This study focuses mainly on material innovations, and more specifically on improved seeds.

In Cameroon, many research bodies (IRAD, IITA, etc.) have included many seed improvement programmes and projects in their activities. Though the innovations ensuing from these research efforts have produced positive impacts at the macro level, the same is not true at the micro level, particularly regarding the performance of farmers (Oehmke and Crawford 1993). The purpose of this chapter is therefore to assess the impact of improved seeds on the productivity of family farms in Cameroon.

This study is divided into four main parts: (1) Data Source and Selected Variables, (2) Analytical Tools, (3) Findings, and (4) Discussion of Findings.

2.2 Data Source and Selected Variables

2.2.1 Data Source

This study specifically focuses on maize. This is because maize is the most widely consumed grain in Cameroon (Gergely 2002). It is grown in virtually all the regions of the country. Furthermore, maize cultivation is an economic, policy and social issue in Cameroon.[2] In addition, statistical records show that the maize sub-sector faces serious performance-related problems, namely:

- low agricultural incomes reflected in the high poverty rate among agricultural households which, according to the National Institute of Statistics (2008), stands at about 60% and
- a serious and ever-growing production deficit (ACDIC 2008).

The study was conducted based on data collected during a survey carried out in August 2007 under the Project to Strengthen Agricultural Research Partnerships in Cameroon designed by the Institute of Agricultural Research for Development. The data base initially comprised 497 family farms located in the West and Centre Regions of Cameroon. These family farms were divided into two groups, namely adopters and non-adopters of improved maize seeds. After carrying out several data processing operations using various criteria, 259 family farms were selected for the study.

2.2.2 Selected Variables

The factors that influence agricultural productivity can be divided into two main groups, namely socio-economic factors and farm assets.

Family farms have many socio-economic features or characteristics whose impact on productivity has been widely studied. Udry (1994) analyses gender efficiency in agricultural production in Burkina Faso and concludes that women-headed family farms are less productive than men-headed family farms. On the other hand, in a study conducted in Kenya, Saito et al. (1994) show that the gender variable has a positive but minor impact on agricultural yields. Findings on the impact of the level of education on agricultural yields are mixed. While some studies confirm that education has a positive and significant impact on agricultural yields (Evenson and Mwabu 1998; Tiamiyu et al. 2009), others prove the contrary (Aguilar 1988). This also applies to farmland. Authors like Bhalla and Roy (1988) and Ntsama (2007) argue that the size of farmland has a positive and significant impact on agricultural productivity. These findings, however, differ from those obtained by authors like Berry and Cline (1979) and Piette (2006).

Farm assets include agricultural equipment (tractors, hoes, etc.), improved or local seeds, insecticides, herbicides, workforce among others (Edwin and Master 2005; Pycroft 2008; Mugisha and Diiro 2010; Okoboi 2010).

The Table 2.1 below presents the variables used in this study.

Table 2.1 Description of model variables

Variables	Description	
Dependent		
rd	Neperian logarithm Agricultural productivity of family farms	Ln (Quantity of maize produced/total area sown with maize)
Independent		
Binary		
Opa	Membership of a farmer organization by the family farm head	1 = member, 0 = not a member
Smat	Marital status of the family farm head	1 = live as a couple, 0 = contrary case
Sex	Sex of family farm head	1 = man, 0 = woman
Sam	Adoption of improved maize seeds	1 = adopt, 0 = do not adopt
Continuous—truncated at zero		
Eagr	Agricultural equipment	Number of tools
Super	Area sown with maize	ha
Mo	Size of workforce	Number of persons employed
Age	Age of family farm head	year

2.3 Analytical Tools

In order to achieve the study objectives, a methodology comprising descriptive and econometric analysis was used to process the sample data. In the first case, the univariate and Bivariate analysis of specific variables were carried out to identify the main characteristics of family farms. The mean difference testing technique was used to conduct a comparative analysis of the average yields obtained by the adopters and non-adopters of improved maize seeds. In the second case, the Blinder-Oaxaca decomposition technique was used to identify and assess the sensitivity of physical agricultural productivity to the adoption of new varieties of maize seeds (Dilling-Hansen et al. 1999; Neuman and Oaxaca 2004; Pycroft 2008).

2.3.1 Econometric Model

$$r\bar{d}^{A} - r\bar{d}^{NA} = \left(\bar{X}^{A} - \bar{X}^{NA}\right)\hat{\beta}^{A} + \bar{X}^{NA}\left(\hat{\beta}^{A} - \hat{\beta}^{NA}\right)$$

where $r\bar{d}^{A} - r\bar{d}^{NA}$ is the average agricultural productivity gap between adopters and non-adopters, $\left(\bar{X}^{A} - \bar{X}^{NA}\right)\hat{\beta}^{A}$ the difference due to the observable characteristics of family farm heads, and $\bar{X}^{NA}\left(\hat{\beta}^{A} - \hat{\beta}^{NA}\right)$ the difference due to the yields of such characteristics. More specifically, rd^{A} and rd^{NA} represent the logarithms of the agricultural yields of adopters and non-adopters respectively. X_i is the vector of the independent variables that can influence productivity, which are presented in Table 2.1. This vector is the same for adopters and non-adopters. Lastly, β^{A}, β^{NA} are coefficients, each measuring the relative contribution of the related independent variable.

It should be pointed out that tests were also conducted to determine selectivity and endogeneity problems; the tests are the maximum likelihood ratio test for selection bias (Hurlin 2002) and the Hausman-Wu-Durbin (or enhanced regression) test for endogeneity bias. Concerning this last test, the financial assistance received by family farm heads and education were retained as instrumental variables. Financial assistance is a potential source of income for farmers. Since improved maize seeds cannot be obtained free of charge, the availability of income can encourage their adoption. Regarding education, it should be noted that the possession of expertise by family farm heads can help in assessing the quality of a proposed product and, hence, encourage its adoption, particularly in terms of the ensuing profitability.

2.4 Findings

2.4.1 Characteristics of Family Farms and Results of Mean Difference Testing

The data analysis showed that there are two types of maize seeds, namely local and improved maize seeds. The statistics obtained indicate that 61.78% of family farm heads use improved maize seeds, while only 38.22% use local varieties. Most of the farmers using improved maize seeds expressed satisfaction with their colour (54.05% of family farm heads), their yield (51.74% of family farm heads) and their quality (53.28% of family farm heads). However, these opinions contrast with those concerning their cycle, taste, size, competitiveness in the market, disease resistance and tolerance to weevils (i.e.50.58%, 50.97%, 52.51%, 55.98%, 65.64% and 69.88% of family farm heads respectively).

It should also be noted that improved seed varieties are derived from diverse sources (IRAD, seed producers, previous harvests, phytosanitary shops, the local market, etc.). This also applies to the methods used to obtain improved seed varieties, of which the main one is purchasing (79.15% of family farm heads).

The mean difference testing results show that though insignificant, the output of the adopters of improved maize seeds is higher than that of non-adopters (see Table 2.3 in Appendix). However, it is necessary to note that these conclusions are assumptions that will be verified using inferential analytical tools.

2.4.2 Existence and Consideration of Selectivity Bias

The findings show that there is a clear selectivity bias for the two productivity equation regimes. The test for independence of equations or maximum likelihood ratio test showed that the coefficients of selection terms or inverse Mills ratio are statistically different from zero at the 1% and 5% thresholds for adopters and non-adopters respectively (see Table 2.4 in Appendix).

This bias was corrected using the Heckman technique (Heckman 1979; Wooldridge 2002; Goulet 2008), which comprises two main stages. At the first stage, the equation of the adoption of improved maize seeds was formulated using the probit model. The estimated coefficients of this equation were used to calculate the inverse Mills ratio for adopters and non-adopters. These ratios were then incorporated into the corresponding agricultural productivity equations. At the second stage, the maximum likelihood (ML) method, instead of the ordinary least squares (OLS) method recommended by Heckman, was used to minimize the inefficiency of the estimators obtained using OLS (Chevassus-Lozza and Galliano 2001). It should be noted that this correction procedure was taken into account when estimating the productivity equations, which accounts for two explanatory components in the agricultural productivity gap decomposition model.

The results of the estimation of agricultural productivity equations are presented again in columns (1) and (2) of Table 2.5 in Appendix.

2.4.3 Agricultural Productivity Gap Between Adopters and Non-adopters

The results of the productivity gap decomposition show that the predicted or estimated values of the average agricultural yield for adopters and non-adopters are 7.286 kg/ha and 6.935 kg/ha respectively, that is, a difference of 0.351 kg/ha (see Table 2.2).

Table 2.2 Results of agricultural productivity gap decomposition

Variables		Predicted values	Gap {(A) – (NA)}
ln (yield)	Adopters (A)	7.286	0.351
	Non-adopters (NA)	6.935	

Gap decomposition (in %) due to											
		Ln (area)	Age	Ln (m.o.)	Ln (equipt.)	Sex	FO	Stat.mat	Inv.mills	Aggregate effects
Difference	of characteristics (E)	–33.9	–8.4	1.7	10.0	0.3	0.1	–0.6	32.7	1.9
	of coefficients (C)	12.0	305.9	–248.1	166.2	24.4	–62.8	59.4	382.8	639.8
	Total aggregate effects {E + C}	*–21.9*	*297.5*	*–246.4*	*176.2*	*24.6*	*–62.7*	*58.8*	*415.5*	*641.7*
	Between the constant model terms (U)									–606.5
Raw differential (R) {E + C + U}										35.1
Adjusted differential (D) {C + U}										33.2
% characteristics (E/R)										5.5
% coefficients (D/R)										94.5

Source: Prepared by the author based on REPARAC 2007 survey data
NB: Positive and negative values respectively represent advantages and disadvantages to adopters

It should also be noted that the constant of the productivity equation of adopters is relatively less than that of non-adopters, which is reflected in the negative value (−606.5%) of the advantages expressed in {U}.

On the whole, the findings show that differences due to the characteristics of individuals and their yields indicate an average agricultural productivity gap between adopters and non-adopters of 1.9% and 639.8% respectively. Furthermore, the immediate consequence of the slight difference in the characteristics of individual of these two groups of farmers is reflected in the narrow gap between the contribution of all model components {E + C + U} and that of the coefficients of the characteristics {C + U} only.

Consideration of the aggregate effects of the characteristics of individuals and their yields {E + C} helps to emphasize that age, number of farm equipment, sex and marital status provide advantages to adopters. This is far from the case for variables such as area sown with maize, workforce and membership of a farmers' organization.

2.5 Discussion of the Findings

The agricultural productivity gap between the adopters and non-adopters of improved maize seeds is positive and is estimated at 0.351 kg/ha. This would mean that, on average, the yield obtained by farmers who adopt improved maize seeds is 1.42 times more than that obtained by those who do not. This finding validates those of many studies conducted by authors like Allogni et al. (2004), Ntsama (2007), Pycroft (2008), Tiamiyu et al. (2009), Mugisha and Diiro (2010), Maruod et al. (2013), Adofu et al. (2013), Kwaku et al. (2014) and Ogunniyi and Kehinde (2015).

It should be pointed out that 5.5% of the gap identified is due to the difference in observable characteristics and 94.5% to the difference in the yields of such characteristics. This implies that if the characteristics of non-adopters were similar to those of adopters, their agricultural yields would increase by 5.5%. Conversely, if the levels of yields of the specific characteristics of adopters were similar to those of non-adopters, their agricultural productivity would drop by 94.5%. The significant contribution of the difference in the yields of specific characteristics in the explanation of this productivity gap could be due to the fact that the use of improved maize seeds often leads to an improvement in the yields of other factors of production. It can therefore be concluded that innovations have a ripple effect on other factors of production.

These findings also show that although there is a productivity gap between the two groups of farmers, it is small compared to the theoretically expected result.[3] This may be due to failure by producers to comply with the conditions associated with obtaining the expected theoretical yield of the said seed varieties. These include the use, method and period of application of fertilizers, the use of herbicides and weeding. In this connection, the sample data show that only 7.5% of adopters carry out weeding on their farms, while 10.6% use herbicides. These low percentages can be explained by the high cost of complying with the relevant guidelines.

2.6 Conclusion

The purpose of this chapter was to assess the impact of improved seeds on the agricultural productivity of family farms in Cameroon. To that end, data collected from 259 family farms which grow only maize were used. The findings show that improved seeds help to increase the productivity of family farms. Some recommendations have therefore been made based on these findings.

It is necessary to facilitate access to some farm inputs (such as pest control products). This is because when they are combined with improved seed varieties, they help to obtain at least the expected theoretical yield of such seed varieties.

Appendix

Table 2.3 Mean difference and mean square deviation testing of farm yields

• Hypotheses and decision rule of mean difference and mean square deviation testing of physical yields

Testing hypotheses

Mean difference testing	*Mean square deviation testing*
H_0: Mean (rd_{NA}) = Mean (rd_A) Diff = 0	H_0: Sd (rd_{NA})/Sd (rd_A) = 1 Ratio = 1
H_1: * Bilateral Mean (rd_{NA}) # Mean (rd_A) Diff # 0	**H_1: * Bilateral** Sd (rd_{NA})/Sd (rd_A) # 1 Ratio # 1
H_1: *Unilateral Mean (rd_{NA}) < Mean (rd_A) Diff < 0 Mean (rd_{NA}) > Mean (rd_A) Diff > 0	**H_1: *Unilateral** Sd (rd_{NA})/Sd(rd_A) < 1 Ratio < 1 Sd (rd_{NA})/Sd (rd_A) > 1 Ratio > 1

Decision rules

With respect to the statistics used

Where $t_{cal} > t_{lu}$, H_0 is therefore rejected	Where $f_{cal} > f_{lu}$, H_0 is therefore rejected
Where $t_{cal} < t_{lu}$, H_0 is therefore accepted	Where $f_{cal} < f_{lu}$, H_0 is therefore accepted

With respect to the computed probability P or α

Where $P < \alpha_{th}$, H_0 is therefore accepted

Where $P > \alpha_{th}$, H_0 is therefore rejected

Source: Prepared by the Author. Mean and Standard Deviation *(Sd)*; α_{th} represents the actual or theoretical significance level. This may be 1%, 5% and 10%; *t* and *f* are the Student t-test and the Fisher t-test, respectively

• Results of the mean difference and mean square deviation testing of farm yields

Variable			Groups		
			Adopters		**Non-adopters**
Ln (farm yield)	Number of observations		160		99
	Mean (*Standard deviation*)		7.286452 (*1.248448*)		6.935129 (*7.021797*)
	Mean difference testing	DL = 257	t = −0.6180	$P_{diff < 0}$ = 0.2686	$P_{diff \# 0}$ = 0.5371
	Mean square deviation testing	DL = 98.159	f = 31.6341	$P_{ratio > 1}$ = 0.0000	$P_{ratio \# 1}$ = 0.0000

Source: REPARAC Survey 2007. t and f are the Student and Fisher-tests

Table 2.4 Selectivity bias testing

Variables	Substantial or productivity equations by group: Column (1) Equation 1: Adopters of improved maize seeds Dependent variable: rd_i^A	Substantial or productivity equations by group: Column (2) Equation 2: Non-adopters of improved maize seeds Dependent variable: rd_i^{NA}	Column (3) Selection or adoption equation Dependent variable: *Sam*
Level of education	0.1529676 (0.70)	0.15359 (0.68)	0.0116556 (0.07)
Area sown with maize	−0.6714924 (4.96)***	−0.1694169 (2.03)**	0.0573592 (0.91)
Age of the family farm head	−0.2059581 (4.05)***	0.0189203 (0.67)	0.0265628 (3.17)***
Membership of a farmers' organization	0.1561473 (0.64)	−0.0685603 (0.28)	−0.0124464 (0.07)
Financial assistance received	−9.556315 (3.94)***	0.868744 (0.88)	1.214225 (3.49)***
Sex of the family farm head	0.6401305 (2.44)**	0.0235961 (0.09)	−0.0810641 (0.45)
Marital status of the family farm head	0.3715151 (1.23)	−0.0034117 (0.01)	−0.0222804 (0.10)
Region of residence	−1.837812 (3.11)***	0.5553122 (1.50)	0.315762 (1.69)*
Distance from the market			0.012048 (0.55)
Size of workforce	0.1483943 (4.03)***	0.0286309 (1.02)	−0. 0128088 (0.62)
Number of agricultural equipment	0.1304015 (2.97)***	0.1509966 (3.32)***	
Use of fertilizers	0.3335662 (1.51)	0.3278144 (1.37)	

Use of insecticides	0.112043	0.0939586	
	(0.63)	(0.50)	
Inverse Mills ratio	9.540569	2.856177	
	(3.81)***	(1.39)	
Constant	7.194357	3.97002	−1.107775
	(12.51)***	(1.52)	(2.50)**
Number of observations	259	259	259
Number of censored observations	99	99	
Number of uncensored observations	160	160	
Wald chi2(k)	52.67	38.20	32.51
Log *(Pseudo)*likelihood	−387.5456	−393.8335	*(−152.51484)*
Prob > chi2	0.0000	0.0003	0.0003

For equation 1: LR test of indep. eqns. (rho = 0): chi2(1) = 8.24 Prob > chi2 = 0.0041
For equation 2: LR test of indep. eqns. (rho = 0): chi2(1) = 5.79 Prob > chi2 = 0.0161

Source: Calculations by author using STATA 11.0. The figures in parentheses represent the absolute value of the z statistic test. The one in parentheses and in italics is the value of the Pseudo-likelihood logarithm. *significance at 10%; **significance at 5%; ***significance at 1%. *"k"* represents the number of independent variables (excluding the constant)

Table 2.5 Results of estimation of farm yield equations

Variables	Column (1) Equation 1: Adopters of improved maize seeds Dependent variable: rd_i^A	Column (2) Equation 2: Non-adopters of improved maize seeds Dependent variable: rd_i^{NA}
Ln (Super) or Area sown with maize	−0.5967682 (5.40)***	−0.3670263 (1.53)
Age of the family farm head	−0.0146481 (1.80)*	−0.0925562 (0.94)
Membership of a farmers' organization	0.0059788 (0.03)	1.485121 (1.26)
Sex of the family farm head	0.2238034 (1.30)	−0.1307447 (0.13)
Marital status of the family farm head	−0.1354805 (0.73)	−0.8896212 (0.48)
Ln (Mo) or Size of workforce	0.1817548 (1.70)*	2.007764 (1.14)
Ln (Eagr) or Agricultural equipment or number of agricultural equipment	0.7936329 (3.67)***	−0.2611372 (0.51)
Inverse Mills ratio	0.7433835 (3.30)***	−4.309337 (0.99)
Constant	5.42132 (11.10)***	11.48678 (1.90)*
Number of observations	160	99
R^2	0.3077	0.0402
F (k, n-k-1)	7.35	3.39
Prob > F	0.0000	0.0019

Source: REPARAC Survey 2007. The figures in parentheses represent the absolute value of the Student (t) test. *significant at 10%; **significant at 5%; ***significant at 1%. *"n"* and *"k"* represent the sample size and number of independent variables of the model (excluding the constant), respectively

Table 2.6 Estimation of expected theoretical average yield of improved maize seeds

		Farm yield (tonne/ha)		
Data sources	Types of seeds	Median values	Mean values	Gap
MINADER (2006)	Local	1.5–2.5	2	2.5
IRAD (2009)	Improved (composite families)	3–6	4.5	

Sources: Calculations made by the author using informations obtained from MINADER (2006) and IRAD (2009)

Notes

1. They represent about 97% of the agricultural labour force (Njonga 2012).
2. From the **sociocultural standpoint**, maize is used for human and animal consumption, as well as for agro-industrial purposes. **At the economic level**, maize production involves a little more than 6 million smallholder farmers in Cameroon (PRP OP Maïs 2008). The activity is very profitable (NEPAD 2004). In addition, demand in maize consumer markets is high and is increasing rapidly in virtually the entire Central African sub-region. Regarding policy, almost all multipurpose projects and programmes of the Ministry of Agriculture and Rural Development (MINADER) have a maize component.
3. **(Computed Yield Gap)** $\Delta rd_{cal} = 432.11 \text{ kg/ha} < \Delta rd_{th} = 2000 \text{ kg/ha}$ **(Theretical Yield Gap)**.

References

Abdel Kwaku, K. B., Donkoh, S. A., & Ayamga, M. (2014). Improved Rice Variety Adoption and its Effects on Farmers' Output in Ghana. *Journal of Development and Agricultural Economics, 6*(6), 242–248.

ACDIC. (2008). Eviter la crise du maïs, Rue CEPER, p. 46. Retrieved from acdic@acdic.net/www.acdic.net.

Adofu, I., Shaibu, S. O., & Yakubu, S. (2013). The Economic Impact of Improved Agricultural Technology on Cassava Productivity in Kogi State of Nigeria. *International Journal of Food and Agricultural Economics, 1*(1), 63–74.

Aguilar, R. (1988). *Efficiency in Production: Theory and Application on Kenyan Smallholders*, Economiska Studier, University of Gotenborg, Sweden.

Allogni, W. N., Coulibaly, O. N., & Honlonkou, A. N. (2004). Impact des nouvelles technologies de la culture de niébé sur le revenu et les dépenses des ménages agricoles au Benin. bulletin de la recherche agronomique du Benin, n 44, 14.

Awotide, B., Diagne, A., & Omonona, B. T. (2012). Impact of Access to Subsidized Certified Improved Rice Seed on Income: Evidence from Rice Farming Households in Nigeria.

Berry, R. A., & Cline, R. W. (1979). *Agrarian Structure and Productivity in Developing Countries*. Baltimore and London: The Johns Hopkins University Press.

Bhalla, S. S., & Roy, P. (1988). Mis-Specification in Farm Productivity Analysis: The Role of Land Quality. *Oxford Economic Papers, 40*(1), 55–73.

Chevassus-Lozza, E., & Galliano, D. (2001), « Les déterminants territoriaux de la compétitivité des firmes agro-alimentaires », Cahiers d'économie et sociologie rurales, n 58–59, 30.

Dilling-Hansen, M., Eriksson, T., Madsen, E. S., & Smith, V. (1999). *The Impact of R & D on Productivity: Evidence from Danish Manufacturing Firms*, p. 23.

ECAM 3. (2007). *Troisième Enquête Camerounaise auprès des Ménages 2007*. Version 1.2 du 13/07/2009. Retrieved from http://www.ilo.org/microdata/index.php.

Edwin, J., & Masters, W. A. (2005). Genetic Improvement and Cocoa Yields in Ghana. *Cambridge University Press, 41*, 491–503.

Evenson, R. E., & Mwabu, G. (1998). *The Effects of Agricultural Extension on Farm Yields in Kenya*. Discussion Paper No 978, Economic Growth Center, Yale University.

Fan, S., Zhang, L., & Zhang, L. (2004). Reforms, Investment and Poverty in Rural China. *Economic Development and Cultural Change, 52*(2), 395–421.

Gergely, N. (2002). *Étude sur l'amélioration de la commercialisation et de la compétitivité des produits agricoles au Cameroun*. FAO.

Goulet, J.-M. (2008). *Processus d'innovation et productivité au Canada: Analyse comparative en fonction de la taille*, mémoire de maîtrise en économie, Université de Sherbrooke, Sherbrooke, p. 109.

Heckman, J. J. (1979). Sample Selection Bias as a Specification Error. *Econometrica, 47*(1), 153–161.

Hurlin, C. (2002). *Econométrie des variables qualitatives, polycopié de cours*. Université d'Orléans.

INS. (2008). troisième enquête camerounaise auprès des ménages: profil de pauvreté en milieu rural au Cameroun en 2007.

Maruod, E. M., Elkhalil, E. B., Elrasheid, E. E., & Ahmed, M. E. (2013). Impact of Improved Seeds on Small Farmers Productivity, Income and Livelihood in Umruwaba locality of North Kordofan, Sudan. *International Journal of Agricultural and Forestry, 3*(6), 203–2018.

Mead, R. W. (2003). A Revisionist View of Chinese Agricultural Productivity. *Contemporary Economic Policy, 21*(1), 117–131.

Mugisha, J., & Diiro, G. (2010). Explaining the Adoption of Improved Maize Varieties and its Effects on Yields Among Smallholder Maize Farmers in Eastern and Central Uganda. *Middle-East Journal of Scientific Research, 5*(1), 6–13.

Neuman, S., & Oaxaca, R. L. (2004). Wage Decompositions with Selectivity-Corrected Wage Equations: A Methodological Note. *Journal of Economic Inequality, 2*, 3–10.

Ntsama Etoundi, S. M. (2007). *Analyse de l'impact de l'innovation sur la productivité agricole: cas du maïs dans la province du Centre-Cameroun*, mémoire de DEA-PTCI, Université de Yaoundé II-Soa, Cameroun, p. 123.

Oehmke, J. F., & Crawford, E. W. (1993). *L'impact de la technologie agricole en Afrique Sub-saharienne: une synthèse des découvertes du symposium*. AMEX International Inc., USAID, p. 38.

Ogunniyi, A., & Kehinde, O. (2015). *Impact of Agricultural Innovation on Improved Livelihood and Productivity Outcomes among Smallholder Farmers in Rural Nigeria*. Maastricht School of Management, Working Paper No 2015/07, pp. 1–23.

Okoboi, G. (2010). *Improved Inputs Use and Productivity in Uganda's Maize Sub-Sector*. Economic Policy Research Centre, Plot 51 Pool Road, Makerere University, p. 32.

Pedelahore, P., & Tchatchoua, R. (2010). *L'innovation est-elle vraiment la solution?: l'exemple du grand Sud Cameroun* (p. 14). Montpellier: ISDA.

Piette, F. (2006). *Les déterminants de la productivité agricole dans le nord-est du Brésil: une investigation sur la relation négative entre la productivité et la taille des fermes* (p. 73). Université de Montréal, département d'économie.

Pycroft, J. (2008). *The Adoption and Productivity of Modern Agricultural Technologies in the Ethiopian Highlands: A Cross-Sectional Analysis of Maize Production in the West Gojam Zone* (p. 20). University of Sussex.

Saito, K., Mekonnen, H., & Spurling, D. (1994). *Raising the Productivity of Women Farmers in Sub-Saharan Africa*. Discussion Paper No 230. World Bank, Washington, DC.

Scarpetta, S., & Tressel, T. (2004). *Boosting Productivity via Innovation and Adoption of New Technologies: Any Role for Labor Market Institutions*. World Bank Policy Research, Working Paper No 3273, pp. 1–31.

Tiamiyu, S. A., Akintola, J. O., & Rahji, M. A. Y. (2009). Technology Adoption and Productivity Difference Among Growers of New Rice for Africa in Savanna Zone of Nigeria. *Tropicultura, 27*(4), 193–197.

Udry, C. (1994). *Gender, Agricultural Production, and the Theory of the Household*. Evanston: Northwestern University.

Wooldridge, J. (2002). *Econometric Analysis of Cross-Section and Panel Data*. Cambridge, MA: The MIT Press.

World Bank. (2008). *World Development Report 2008: Agriculture for Development*. Washington, DC: World Bank.

Check for updates

3

Breaking the Traditional Trap: Assessing Drivers of Modern Technology Adoption by Smallholder Farmers in Hurungwe District, Zimbabwe

Carren Pindiriri

3.1 Introduction

The importance of adopting modern technology in agriculture, especially in a changing climate, cannot be underestimated in Africa. Many studies (Kijima et al. 2008; Mendola 2007; Liu and Wang 2005; de Janvry and Sadoulet 2002; Xu and Jeffrey 1998) demonstrate that agricultural modernisation increases productivity. Diffusion of modern agricultural technologies also enhances sustainable development through poverty elimination (Kassie et al. 2011; Suri 2011; Duflo et al. 2008). Concurring with these researchers, Boniphace et al. (2015) identify lack of agricultural investment and insufficient usage of modern technologies as some of the factors impeding agriculture growth in Africa. Moreover, in the 2016 African Development Bank's (AfDB) strategic plan, agricultural development through improved technologies is critical in promoting one of the bank's high 5s, namely, feeding Africa. Despite its importance, the

C. Pindiriri (✉)
University of Zimbabwe, Harare, Zimbabwe

A. Shimeles et al. (eds.), *Building a Resilient and Sustainable Agriculture in Sub-Saharan Africa*, https://doi.org/10.1007/978-3-319-76222-7_3

uptake of modern agricultural technologies has, however, remained very low in sub-Saharan Africa (Langat et al. 2013; Gollin et al. 2005).

For instance, Gollin et al. (2005) reveal that, in 2000, only 17% of the area planted for maize had modern maize varieties in sub-Saharan Africa compared to 57% in Latin America and the Caribbean. This low uptake of agricultural technologies is a cause for concern in Africa where food security is severely threatened by the changing climate. While the adoption of modern agricultural technologies has been identified as the main driver of green revolution in Asian countries (Ravallion and Chen 2004), it still remains a puzzle why the adoption rates of these agricultural technologies have remained low in sub-Saharan Africa (Matsumoto et al. 2013; World Bank 2008). Mkandawire and Matlosa (1994) even query why the green revolution which transformed agriculture in Asia and Europe failed to achieve similar results in sub-Saharan Africa. This suggests that the African continent has been trapped in the traditional production methods and has therefore remained the world's greatest laggard in agricultural technological growth (Boko et al. 2007).

Despite experiencing a decline, agriculture remains the backbone of the Zimbabwean economy, with maize production anchoring food security in the country. Maize is produced in all provinces of the country, but the largest share of maize output is from Mashonaland provinces. Hurungwe, in Mashonaland West, is the largest district and one of the major maize-producing districts in the country. The district has the potential to significantly improve food security because of its favourable climatic conditions. It is therefore vital to take advantage of the district's potential in improving national maize production by enhancing farmers' productive capacity through various ways, which include the promotion of modern technologies in agriculture. It is in this view that this chapter scrutinises the drivers of modern technology adoption in Hurungwe. The main objectives of the chapter are therefore to: (1) measure the agricultural technological gap for Hurungwe farmers and (2) examine the drivers of modern agricultural technology adoption in Hurungwe. An appreciation of the drivers of technology adoption in the district helps in: (1) identifying the characteristics of adopters and predicting adoption rates, (2) identifying policy targets for improving adoption rates, (3) enhancing Sustainable Development Goals (SDGs) and (4) developing marketing strategies for new technologies (Oster and Thornton 2012).

Although a substantial amount of work has been done on the determinants of agricultural technology adoption across the world, very little has been done in Zimbabwe. Agricultural technology adoption has generally been regarded as a very slow process whose many aspects have continued to be poorly understood (Simtowe et al. 2011 and Diagne and Demont 2007). Studies that have considered Zimbabwe in agricultural technology adoption were mainly done for sub-Saharan Africa (Muzari et al. 2012; Boko et al. 2007; Mkandawire and Matlosa 1994). These studies nevertheless fall short of proper methodological approaches to the exploration of the drivers of technology adoption in agriculture as they overlook non-exposure and selection biases prevalent in classical technology adoption models such as probit and logit applied by many researchers (Fadare et al. 2014; Hailu et al. 2014; Zivanemoyo and Mukarati 2013; Ayoola 2012).

This chapter is therefore expected to add to the list of existing literature on the drivers of modern technology adoption in agriculture. It adds new literature on agricultural technology adoption in Zimbabwe in the following ways: first, the selection of a study area which has never been investigated in the area of determinants of agricultural technology adoption helps in unmasking the cloaked. Second, focusing on a particular district with identical culture among farmers helps in avoiding misleading estimators from national-based models which provide an average coefficient for heterogeneous areas. Third, the chapter applies a non-classical adoption technique (average treatment effects) to remedy the problems resulting from non-exposure and selection biases. Despite being used in a number of countries (Simtowe et al. 2011; Diagne and Demont 2007), the average treatment effects (ATE) technique has never been applied to study the determinants of technology adoption by famers in Zimbabwe. The chapter therefore extends the application of this technique to Zimbabwean farmers.

3.2 Literature Survey

The technology adoption curve has for long been regarded by sociologists and marketers as a normative and descriptive model to decision-making just like the product life cycle (Anderson and Zeithaml 1984; Midgley 1977; Rogers 1962). Adopting a new technology in the traditional school is associated with different categories of adopters, some being innovators (immediate adopters of a new technology) and others being laggards (last group to adopt a new technology). The product life cycle theory, thus, regards young people as innovators and early adopters while the elderly are regarded as laggards. From the 1980s, researchers began to question the applicability of the traditional product life cycle because of its rigid assumption regarding the 'S' shape which could not match empirical data in many cases. In this view, researchers such as Lambkin and Day (1989), Bayus (1988) started to extend the traditional product life cycle through diffusion models. Foster and Rosenzweig (2010) argue that a farmer makes a decision to adopt technology if the technology is expected to stay profitable and if it is available and affordable. In this context of profit maximisation, Sunding and Zilberman (2000) and Pingali et al. (1987) further demonstrate how farm size restricts technology adoption in a profit-maximising problem. However, farmers may choose to adopt a technology in anticipation of future benefits even if it is not currently profitable (Smale et al. 1995).

Many studies have, however, been done to investigate the factors explaining agricultural technology adoption rates despite having scanty literature in Zimbabwe. Plenty of literature on agricultural technology adoption is available in many countries, with developing countries recently contributing a significant share of the literature (Diagne and Demont 2007). These empirical findings reveal that adoption of agricultural technologies relies on farmers' perceptions about the technology (Rogers 1962) and further classify drivers of technology adoption into farmers' socio-demographic factors, institutional forces and farmers' economic status (Doss et al. 2003). In some cases, drivers of agricultural technology adoption have been categorised as market motivations (profit and risk), bio-physical drivers and farmers' preferences (Pattanayak et al. 2003). Technology adoption has been defined either as continuous or discrete variable in these studies (Doss et al. 2003). However, in many cases it has been considered as a discrete variable because of the complexities involved when measuring it as a continuous variable, especially in African agriculture where farmers rarely keep records of input purchases. The evidence produced by Pattanayak et al. 2003 meta-analysis indicates that over 95% of the 32 reviewed studies on agro-forestry technology adoption have measured technology adoption as a discrete variable and applied either the probit, logit or linear probability models.

Many more studies on agricultural technology adoption have applied discrete dependent variable models and identified the following drivers: age of the farmer, farm size, exposure to technology, access to credit, farmer's education, access to extension services, gender, household size, income, farming experience, neighbourhood, climatic conditions and agricultural training among others (Boniphace et al. 2015; Fadare et al. 2014; Hailu et al. 2014; Langat et al. 2013; Uaiene et al. 2009; Dimara and Skuras 2003). Although there is general consensus among the researchers with regard to the effect of all the other identified factors on technology adoption, the effect of farm size on the farmer's decision to adopt modern technology has remained unsettled. Akudugu et al. (2012) argue that the effect of farm size on adoption of agricultural technologies can either be positive, harmful or impartial. For example, Langat et al. (2013), Uaiene et al. (2009) and Feder et al. (1985) established a positive association between farm size and technology adoption while Harper et al. (1990) found farm size to have a negative effect on agricultural technology adoption. Other studies even established a neutral relationship between farm size and technology adoption (Fadare et al. 2014; Reimer and Fisher 2014).

Despite its extensive nature, literature on drivers of technology adoption in agriculture has its own drawbacks. First, a majority of the studies mainly focussed on drivers of adoption of hybrid seeds (Boniphace et al. 2015; Fadare et al. 2014; Langat et al. 2013; Zivanemoyo and Mukarati 2013; Simtowe et al. 2011) while overlooking farm mechanisation. Only few studies, for example, Akudugu et al. (2012), Uaiene et al. (2009) and Dimara and Skuras (2003) considered farm mechanisation as an equally important type of technology adoption worth to be investigated in farming households. The transformation of communal farmers from subsistence entities into business entities through modernisation of agricultural production systems is essential for improving food security in Africa. An investigation of the determinants of farm mechanisation is therefore critical in the African continent where most of the smallholder farmers are trapped in the traditional production systems.

Second, most of the studies (over 95%) have applied the classical adoption models, namely logit and probit. In most cases, the estimated parameters of these models tend to underestimate the true population parameters of adoption determinants due to selection and non-exposure biases inherent in discrete adoption models (Diagne and Demont 2007). Very few researchers have, however, recently turned their attention to the use of methods that remedy these biases. For example, Simtowe et al. (2011) applied a programme evaluation technique to investigate the determinants of adoption of improved Pigeon pea varieties in Tanzania. The use of improved methodologies helps in avoiding misleading policy recommendations.

Third, despite being one of the major maize-producing countries in Africa, no attempt has been made to empirically examine drivers of technology adoption by maize farmers in Zimbabwe. This is a huge motivation for this study. Hurungwe District has excellent climatic conditions for maize production and remains the main maize-producing district in Zimbabwe. Modernisation of agriculture in the district will go a long way in feeding Zimbabwe and other sub-Saharan African countries.

3.3 Methodology and Data Issues

Classical economists argue that farmers can only adopt new technology if they are exposed to it (Foster and Rosenzweig 2010). Awareness is therefore a necessary condition for adopting a new technology. However, when a new technology is introduced, farmers may not be universally exposed to it, as a result the observed sample parameter may not be a consistent estimator for the true population parameter. Diagne and Demont (2007) argue that applying classical models of adoption when the target population is not universally exposed to the new technology may result in a non-exposure bias which produces biased and inconsistent estimators for population adoption rates. In addition, Simtowe et al. (2011) show that farmers' exposure to a new technology is non-random since extension workers may target farmers with higher probability of adopting or farmers may get exposed through their self-interests. Exposure to a new technology therefore suffers from selection bias signifying a non-linear association between exposure to and adoption of a new technology. It is against this background that this chapter evaluated the drivers of technology adoption using a programme evaluation methodological approach as in Wooldridge (2002), Diagne and Demont (2007) and Simtowe et al. (2011).

Consider i as indexing farmers and $\boldsymbol{E_i}$ as a treatment indicator, equal to 1 if the farmer is exposed to agricultural technology, that is, if the farmer is treated and equal to 0 if the farmer is not exposed to agricultural technology (not treated). Farmers exposed to agricultural technology were referred to as the 'treated' while those not exposed to technology were the 'untreated'. A farmer was said to have adopted modern agricultural technology if he/she had adopted at least one of the following as an independent farmer: hybrid or improved seed variety, a tractor, a pump, a harvester, a planter, a generator, modern irrigation equipment or modern weather forecasting equipment. Further, consider $\boldsymbol{\pi_{i0}}$ and $\boldsymbol{\pi_{i1}}$ to be the potential adoption outcomes that would occur when a farmer is not treated ($\boldsymbol{E_i}$ = 0) and when a farmer is treated ($\boldsymbol{E_i}$ = 1), respectively. Either $\boldsymbol{\pi_{i0}}$ or $\boldsymbol{\pi_{i1}}$ is observable but not both. For example, we can only observe that an untreated farmer has not adopted agricultural technology but we cannot certainly deduce what would have been the outcome if this farmer was exposed to the technology. The inference is therefore counterfactual, an adoption outcome that would have happened if the farmer was exposed to technology. In other words, the impact of exposure on technology adoption on the same farmer cannot be measured and this is referred to as the problem of missing data (Dimara and Skuras 2003).

The agricultural technology adoption outcome for the ith farmer was therefore given as:

$$\pi_i = \boldsymbol{E}_i \pi_{i1} + \left(1 - \boldsymbol{E}_i\right) \pi_{i0} \tag{3.1}$$

Equation (3.1) can equally be expressed as:

$$\pi_i = \pi_{i0} + \left(\pi_{i1} - \pi_{i0}\right) \boldsymbol{E}_i = \alpha_i + \beta_i \boldsymbol{E}_i \tag{3.2}$$

where $\boldsymbol{\alpha_i} = \boldsymbol{\pi_{i0}}$ and $\boldsymbol{\beta_i} = \boldsymbol{\pi_{i1}} - \boldsymbol{\pi_{i0}}$ are the intercept and the treatment effect for the ith farmer, respectively. Since only one of the components of $\boldsymbol{\beta_i}$ is observable, the treatment effect ($\boldsymbol{\beta_i}$) is unidentified but we can identify useful measures namely: (1) ATE which averages the entire population of the farmer treatment effects or averages $\boldsymbol{\beta}_i$ over all the sampled farmers, (2) the average adoption outcome of the treated or farmers exposed to technology (ATET) which averages $\boldsymbol{\beta}_i$ over a sub-set of farmers exposed to technology and (3) the average adoption outcome of the untreated farmers (ATENT) which averages $\boldsymbol{\beta}_i$ over a sub-set of farmers not exposed to technology. The three measures of treatment effect are measured as:

$$\text{ATE} = E(\beta_i) = E(\pi_{i1} - \pi_{i0}) = \beta = E(\pi_1 - \pi_0) \tag{3.3}$$

$$\text{ATET} = E\left[\beta_i | \boldsymbol{E}_i = 1\right] = E\left[(\pi_{i1} - \pi_{i0}) | \boldsymbol{E}_i = 1\right] \tag{3.4}$$

$$\text{ATENT} = E\left[\beta_i | \boldsymbol{E}_i = 0\right] = E\left[(\pi_{i1} - \pi_{i0}) | \boldsymbol{E}_i = 0\right] \tag{3.5}$$

Since exposure to technology is usually a necessary condition for technology adoption, it implies that $\boldsymbol{\pi}_0 = 0$ and ATE = $E(\boldsymbol{\pi}_1)$. The difference between ATE and ATET is called the population selection bias (Wooldridge 2002). In order to produce unbiased and consistent estimators for ATE and ATET, there is need to control for this population selection bias (Diagne and Demont 2007). Farmers who adopt agricultural technologies become exposed to those technologies, hence, the need to correct the likely problem of endogeneity where exposure to technology is also determined within the system. An endogenous treatment of binary outcomes was hence applied in this chapter.

Variables used in this chapter came from the reviewed literature. The two endogenous binary variables are (1) agricultural technology adoption ($\boldsymbol{\pi}$) by a farmer which took a value of 1 if the farmer had adopted any agricultural technology (either mechanisation or hybrid seeds) as an independent farmer and 0 otherwise and (2) farmer's exposure to agricultural technologies ($\boldsymbol{E}$) which took a value of 1 if the farmer was exposed to any agricultural technology and zero otherwise. The modern technologies considered in this chapter include (1) mechanical which consists of tractors, harvesters, planters, irrigation equipment such as water pumps and generators and (2) biological and geographical which consist of improved seed varieties and forecasting methods. The two endogenous variables, $\boldsymbol{\pi}$ and $\boldsymbol{E}$, are determined by vectors of covariates, $\boldsymbol{X}$ and $\boldsymbol{Z}$, respectively. In summary, technology adoption and its drivers can be estimated from random vectors, ($\boldsymbol{\pi}_i$, $\boldsymbol{E}_i$, $\boldsymbol{X}_i$, $\boldsymbol{Z}_i$) for $i = 1 \ldots n$. In order to estimate ATE, the treatment condition ($\boldsymbol{E}$) is assumed to be independent of the possible adoption outcomes, $\boldsymbol{\pi}_0$ and $\boldsymbol{\pi}_1$, conditional on a vector of covariates $\boldsymbol{Z}$ that explain exposure, that is, $Prob[\boldsymbol{\pi}_s = 1 \mid \boldsymbol{E}, \boldsymbol{Z}] = Prob[\boldsymbol{\pi}_s = 1 \mid \boldsymbol{Z}]$ for $s = 0, 1$. This is referred to as the conditional independence axiom (Wooldridge 2002). The population mean technology adoption conditional on vector $\boldsymbol{X}$ is given as:

$$\mathrm{ATE}(X) = E\left[\pi_1 = 1 | X\right] \tag{3.6}$$

One way of estimating the ATE parameters is to interact $\boldsymbol{E}$ with covariates and then apply the usual parametric regression-based approaches. The second way, which was used in this chapter, is the application of a two-stage estimation technique. First, a propensity score was generated through regressing treatment, $\boldsymbol{E}$, on its covariates vector, $\boldsymbol{Z}$, that is, $Prob[\boldsymbol{E} = 1 \mid \boldsymbol{Z}] = Prob(\boldsymbol{Z})$. Second, ATE was estimated by parametric techniques. With non-parametric approach, the conditional independence assumption is extended to include the independence of possible adoption from the drivers of treatment ($\boldsymbol{Z}$) conditional on vector $\boldsymbol{X}$, that is, $Prob[\boldsymbol{\pi}_1 = 1 \mid \boldsymbol{X}, \boldsymbol{Z}] = Prob[\boldsymbol{\pi}_1 = 1 \mid \boldsymbol{X}]$. When using the parametric approach as done in this chapter, the conditional independence assumption allows us to estimate technology adoption and its drivers from the treated sub-sample only through the following specification:

$$E\left[\pi | \boldsymbol{X}, \boldsymbol{E} = 1\right] = f\left(\boldsymbol{X}, \lambda\right) \tag{3.7}$$

where f is an identified linear or non-linear function of a vector of explanatory variables $\boldsymbol{X}$ and unknown parameter vector $\boldsymbol{\lambda}$ to be estimated. The estimated equation was then used to compute the predicted values which were then used to estimate the ATE and ATET for the whole sample and treated sub-sample, respectively. The farmers' technology adoption gap (GAP) is the deviation of ATE from the joint exposure and adoption parameter (JEA).

Technology adoption literature identifies many factors explaining farmers' decision to adopt new technologies and their exposure to agricultural technologies. Table 3.1 provides a summary of the determinants of exposure to and adoption of agricultural technologies, that is, the variables in vectors $\boldsymbol{Z}$ and $\boldsymbol{X}$. The variables, age (in years), farm size (in hectares), extension services (number of contacts per year), income (dollars), farming experience (in years), education (in completed years), urbanity (in years), household size (number of members) and bread winner urbanity (in years) were measured as continuous variables while the rest were measured as dichotomous variables (see Table 3.1).

The data used in this chapter were collected using a questionnaire from a sample of 411 farmers subjected to an experiment in Hurungwe. A multistage sampling procedure was carried out. First, wards were stratified according to ecological zones and one ward was then randomly selected from each ecological zone (regions IIA, III and IV). Only ecological region V was disregarded because the region is set aside for wildlife management. Each selected ward was proportionally represented in terms of the sampling units. Enumeration areas (EAs), as demarcated by the Zimbabwe Statistical Agency (Zimstat) in 2012, within each ward, were then randomly selected and a census was carried out within the selected six EAs. Farmers not exposed to technology were considered to constitute a control group.

Table 3.1 Determinants of exposure and technology adoption

Adoption determinants (*X*)	Expected sign	Exposure determinants (*Z*)	Expected sign
Exposure to technology (*E*)	+	Extension services	+
Age of the farmer	+/–	Age of the farmer	+/–
Farm size	+/–	Farmer's education	+
Credit (=1 for access to)	+	Urbanity	+
Farmer's education	+	Gender (=1 for male)	+
Extension services	+	Training (=1 for trained)	+
Gender (=1 for male)	+	Knowledge source	+
Belief (=1 for traditional)	–	Farmer's experience	+
Income	+		
Farming experience	+		
Weather (=1 for wet)	+		
Perception (=1 for +ve)	+		
Training (=1 for agricultural trained)	+		
Parent belief (1 if tradition)	–		

3.4 Results and Discussion

The findings show that 76.9% of the interviewed farmers were male and 80.3% were exposed to at least one type of agricultural technology. Despite many farmers being exposed to agricultural technologies in the district, only 30.2% of the farmers adopted modern agricultural technologies. Table 3.3 presents descriptive statistics of the sampled farmers categorised according to their adoption status of agricultural technologies. The statistics demonstrate that the difference between the proportion of adopters and the proportion of non-adopters in wet ecological zones and dry zones is statistically insignificant. Similarly, the difference between the average farm size of adopters and that of non-adopters is statistically insignificant.

The statistics, however, display significant differences between adopters and non-adopters in terms of gender representation, education, beliefs, credit access, technology exposure, technologies perceptions, knowledge sources, parents' beliefs, age, extension contacts, farming experience and incomes. The male to female ratio is bigger in the sub-sample of farmers who adopted agricultural technologies. With regard to age, the statistics in Table 3.3 concur with the theoretical supposition that when a new technology is introduced, older farmers take time to adopt it as they are reluctant to disturb their tradition. The average age in the adopters' category (41 years) is significantly less than the average age of non-adopters (45 years). Similarly, the average farming experience for adopters (14 years) is smaller than that of non-adopters (17 years). The proportion of farmers who believe in tradition is bigger in the non-adopters' sub-sample than in the adopters'. Furthermore, the statistics show an ordinarily larger percentage of farmers in the non-adopters sub-sample whose parents were traditionalists. In concurrence with the larger proportion of traditionalists in the non-adopters' group, the percentage of farmers who perceive modern technology to be better than traditional technologies is larger in the adopters' sub-sample. These statistics, generally, point to an important implication that farmers who are bonded to their tradition find it difficult to adopt modern technologies.

The statistics in Table 3.2 indicate that average education is significantly higher for adopters (11 years) than for non-adopters (8 years). Similarly, the percentage of agriculturally trained farmers is larger in the

Table 3.2 Farmers' characteristics according to their adoption status

Characteristic	Adopters (*N* = 124)	Non-adopters (*N* = 287)	Total (*N* = 411)	Difference
Proportion of male farmers (%)	87.1	72.3	76.9	14.6***
Proportion of farmers in wet zones (%)	62.1	56.4	58.2	5.7
Proportion of traditionalists (%)	35.5	62.0	54.0	−26.5***
Proportion of agric trained farmers (%)	33.9	2.8	12.2	31.1***
Proportion of farmers with credit access (%)	75.8	15.7	33.8	60.1***
Farmers exposed to modern technology (%)	89.5	76.3	80.3	13.2***
Farmers saying modern tech is better (%)	98.4	54.4	67.6	44.0***
Farmers with traditional parents (%)	47.6	80.1	70.3	−32.6***
Farmers with modern knowledge sources (%)	79.0	22.3	39.4	56.7***
Average farm size (hectares)	8.70	8.76	8.74	−0.1
Average age of farmers (years)	41.0	45.0	44.0	−4.0***
Average education of farmers (years)	11.0	8.0	9.0	3.0***
Average extension contacts (number per year)	15.0	6.0	10.0	9.0***
Average farming experience (years)	14.0	17.0	16.0	−3.0**
Average yearly income (dollars)	6407	1654	3088	4753***

***, ** and * indicate that the difference between adopters and non-adopters is statistically significant at 1%, 5% and 10% level, respectively

adopters' category. Although the findings show that only 12.2% of farmers were agriculturally trained, 33.9% of the adopters were agriculturally trained and only 2.8% of the non-adopters were agriculturally trained. Moreover, the average extension contact visits for adopters are significantly more than that of non-adopters. Increased extension visits are associated with increased farmer education, hence increased exposure to agricultural technologies. Likewise, 79% of the adopters have access to modern knowledge sources compared to only 22.3% of the non-adopters. These findings suggest a positive association between education and the decision to adopt agricultural technologies. Although access to credit among farmers is low (33.8%), the difference between adopters and non-adopters is statistically significant, a contradiction to Simtowe et al. (2011) findings. About 75.8% of the adopters had access to credit while only 15.7% of the non-adopters had access to credit. The average income of adopters was significantly larger than that of non-adopters. This implies a positive association between technology adoption and exposure to technology, access to credit and income.

The findings from probit regressions presented in Table 3.3 show that only extension contacts and urbanity of the farmer explain awareness of agricultural technologies in Hurungwe District. The coefficients of the two variables are statistically significant at 10% level. The three probit models presented in columns (1)–(3) show a high degree of consistence of the factors explaining farmers' exposure to agricultural technologies. In all models, an increase in the number of extension contacts and urban experience increase the probability of the farmer's exposure to agricultural technologies. The main implication of this finding is that intensifying extension services in smallholder farms will improve farmers' awareness to improved agricultural technologies. Similar findings were established by Hailu et al. (2014) and Simtowe et al. (2011).

Table 3.3 Determinants of the probability of exposure to agricultural technologies

	(1)	(2)	(3)
Variables	Exposure	Exposure	Exposure
Gender of farmer	0.248	0.239	0.258
	(0.189)	(0.187)	(0.177)
Education of farmer	0.0158	0.00932	0.00622
	(0.0177)	(0.0205)	(0.0178)
Farmer's experience	0.00939*	0.00290	
	(0.00555)	(0.00956)	
Urbanity	0.0325**	0.0279*	0.0272*
	(0.0158)	(0.0168)	(0.0166)
Extension	0.0272*	0.0286*	0.0289*
	(0.0155)	(0.0155)	(0.0154)
Age		0.00451	0.00594*
		(0.00574)	(0.00331)
Wald Chi-square	143***	145.7***	145.9***
Observations	411	411	411

Standard errors in parentheses
***$p < 0.01$, **$p < 0.05$, *$p < 0.1$

Table 3.4 Summary of adoption rates of agricultural technologies in Hurungwe

Adoption is the dependent variable			
Variable	Coefficient	PO means	Coefficient
ATE (exposure 1 vs. 0)	0.429***	PO mean (exposure 0)	0.032***
ATET (exposure 1 vs. 0)	0.330***	PO mean (exposure 0)	0.000
PO mean (exposure 1)	0.460***	PO mean (exposure 0)	0.032***

PO stands for potential outcome. ***, ** and * indicate that the coefficient is statistically significant at 1, 5 and 10% level, respectively

The results presented in Table 3.4 reveal that exposure to technology causes adoption rates of agricultural technologies to increase by an average of 42.9% from the average of 3.2% of farmers not exposed to technologies. This is referred to as the ATE. The findings further show that among the exposed farmers, exposure causes adoption rates to increase by an average of 33% from the average of 0% that would have occurred if the farmers had not been exposed to agricultural technologies. This is the ATET. Potential outcome (PO) means indicate that the average rate of adoption for exposed farmers is 46% while for non-exposed farmers is only 3.2%. The actual adoption rate is 30.2% and the population adoption gap emanating from farmers' incomplete exposure to agricultural technologies is 12.7%. There is potential to improve adoption rates of agricultural technologies by 12.7% in Hurungwe through making farmers aware of the existing technologies. The population selection bias measured by the difference between ATE and ATET was found to be 9.9% and statistically significant at 1% level. The statistically significant selection bias demonstrates that the probability of technology adoption for a farmer in the treated group is different from the probability of technology adoption for a farmer randomly selected from the population. This therefore justifies the application of treatment effects in this chapter.

A two-step estimation technique was applied to a probit model with treatment effects in examining the drivers of agricultural technology adoption in Hurungwe. The estimated models are presented in Table 3.5 from column (1) to column (3). Coefficients which are statistically significant in the first model (column 1) remain statistically significant in the other two models signifying a reasonable degree of reliability. The findings show that factors which include education of the farmer, agricultural training, access to credit, per capita income, perception about

Table 3.5 Drivers of technology adoption from a two-step treatment effects probit model

	(1)	(2)	(3)
Variables	Adoption	Adoption	Adoption
Age	0.00001		
	(0.00274)		
Gender (1 for males)	−0.029		
	(0.054)		
Education	0.014**	0.012*	0.011*
	(0.007)	(0.007)	(0.006)
Experience	0.003	0.002	0.002
	(0.003)	(0.002)	(0.002)
Training (1 for agriculturally trained)	0.161**	0.161**	0.164**
	(0.072)	(0.075)	(0.071)
Extension	0.0002	0.0002	
	(0.0011)	(0.0011)	
Credit (1 for farmers with credit access)	0.326***	0.326***	0.308***
	(0.044)	(0.043)	(0.043)
Farm size	−0.0008	−0.0004	
	(0.0031)	(0.0031)	
Per capita income	0.001***	0.001***	0.001***
	(0.0003)	(0.0002)	(0.0002)
Perception (1 for modern believers)	0.166***	0.166***	0.158***
	(0.0422)	(0.0420)	(0.0416)
Parent belief (1 for traditional parents)	−0.107***	−0.108***	−0.100***
	(0.040)	(0.040)	(0.0386)
Constant	−0.549**	−0.598**	−0.425*
	(0.235)	(0.237)	(0.225)
Wald Chi-square	330.7***	316.2***	346***
Observations	409	409	409

Standard errors in parentheses
***$p < 0.01$, **$p < 0.05$, *$p < 0.1$

modern technologies and beliefs of farmer's parents explain farmers' decision to adopt modern agricultural technologies. The coefficients of these factors were found to be statistically significant at 10%, 5% or 1% level. While a number of studies (Hailu et al. 2014; Langat et al. 2013; Akudugu et al. 2012; Uaiene et al. 2009) established an association between a farmer's decision to adopt modern technology and his/her age, sex, experience, belief, extension services and farm size, this chapter established otherwise. Findings in this study are however in line with Simtowe et al. (2011) who found no association between adoption of improved Pigeon pea varieties and age and gender of farmers in Tanzania. Uaiene et al. (2009) also found no association between farm size and adoption of mechanical agricultural technologies in Mozambique.

Education of the farmer and agricultural training are positively associated with the probability of technology adoption. Improved education and training increase the farmer's propensity to adopt agricultural technologies by 1% and 16%, respectively. Formal specialised training in agriculture has more impact on farmers' decision to adopt modern technologies compared to just formal education. Similar findings were established by Boniphace et al. (2015) in Tanzania, Kassie et al. (2011) and Kijima et al. (2008) in Uganda and Fadare et al. (2014) in Nigeria. The main implication of this finding is that increased education and training of smallholder farmers can improve adoption rates of agricultural technologies, hence improving food security for many African countries. Whereas education and training increase the farmer's probability to adopt agricultural technologies, bondage in tradition has a negative influence on farmers' decisions to adopt modern technologies. Farmers with traditionally bonded parents have more than 10% lower probability of adopting modern agricultural technologies. This is also buttressed by the effect of perception on technology adoption. Farmers who perceive modern technologies as better that traditional technologies have a 16% to 17% higher propensity to adopt modern technologies in agriculture.

The results further show that increased incomes and access to credit increase the farmer's probability to adopt modern technologies by 0.1% and 30–33%, respectively. Affordability of a given technology is critical when a farmer makes the final decision on whether to adopt a given technology or not. Hailu et al. (2014) and Uaiene et al. (2009) established similar results in Northern Ethiopia and Mozambique, respectively. In concurrence with these studies, Feder et al. (1985) argue that credit access constraints are often cited as the main reason why farmers fail to adopt modern agricultural technologies. Agricultural equipment is quite expensive to most smallholder farmers and in this regard access to credit becomes crucial in the technology adoption decision-making process. As argued by Feder et al. (1985), insufficient accumulated savings by smallholder farmers prevent them from investing in modern agricultural technologies; hence, availability and access to credit may close this gap.

3.5 Conclusion and Policy Implications

This study measured the agricultural technology gap and examined the drivers of modern technology adoption by maize farmers in Hurungwe, Zimbabwe. The study findings verify the presence of sample selection bias, hence the importance of using treatment effects. ATE results demonstrate that exposure to technology causes adoption rates of agricultural technologies to increase by an average of 42.9% from the average of 3.2% of farmers not exposed to technologies. The ATET findings further show that among the exposed farmers, exposure causes adoption rates to increase by an average of 33% from the average of 0% that would have occurred if the farmers had not been exposed to agricultural technologies. The population technology adoption gap caused by non-exposure of farmers to agricultural technologies in Hurungwe is 12.7%.

Extension services and urbanity were found to be the main determinants of exposure while access to credit, income, education, training and positive perception about modern technologies were found to increase the farmer's propensity to adopt modern agricultural technologies. But farmers with traditionally bonded parents were found to have a lower propensity to adopt modern agricultural technologies. These findings point to important policy implications. First, there is potential to improve agricultural technology adoption rates through improving farmers' exposure to technologies via various media such as radios, television and extension services among others. As revealed in this chapter, education and formal training also increase farmers' propensity to modernise their production systems.

Second, the results show that improved farmers' incomes and access to credit increase their propensity to adopt modern agricultural technologies. Access to credit has a significant effect on the farmer's decision to adopt modern technologies. The major implication of this finding is that financial inclusion for farmers is critical for modernising African agriculture. In many African countries, farmers face difficulties in accessing credit, leading to lack of investment in new agricultural technologies. This finding therefore suggests that financial inclusion through establishment of rural financial institutions can significantly aid modernisation of Zimbabwean agriculture.

In conclusion, the findings generally point to the need for improving farmers' access to credit, reducing technological information asymmetry amongst farmers and intensifying education and agricultural training for farmers.

References

Akudugu, A. M., Guo, E., & Dadzie, K. S. (2012). Adoption of Modern Agricultural Production Technologies by Farm Households in Ghana: What Factors Influence Their Decisions? *Journal of Biology, Agriculture and Healthcare, 2*(3), 1–13.

Anderson, C. R., & Zeithaml, C. P. (1984). Stage of the Product Life Cycle, Business Strategy, and Business Performance. *Academy of Management Journal, 27*(1), 5–24.

Ayoola, B. J. (2012). Socio-Economic Determinants of the Adoption of Yam Minisett Technology in the Middle Belt Region of Nigeria. *Journal of Agricultural Science, 4*(6), 215–222.

Bayus, B. L. (1988). Accelerating the Durable Replacement Cycle with Marketing Mix Variables. *Journal of Product Innovation Management, 5*(3), 216–226.

Boko, M., Niang, I., Nyong, A., Vogel, C., Githeko, A., Medany, M., Osman-Elasha, B., Tabo, R., & Yanda, P. (2007). *Africa. Climate Change 2007.* Impacts, Adaptation and Vulnerability. Contribution of Working Group II to the Fourth Assessment Report of the Intergovernmental Panel on Climate Change, Parry, M. L., Canziani, O. F., Palutikof, J. P., van der Linden, P. J. & Hanson, C. E. Eds., Cambridge University Press, Cambridge, 433–467.

Boniphace, N. S., Fengying, N., & Chen, F. (2015). An Analysis of Smallholder Farmers' Socio-Economic Determinants for Inputs Use: A Case of Major Rice Producing Regions in Tanzania. *Russian Journal of Agricultural and Socio-Economic Sciences, 38*(2), 41–55.

De Janvry, A., & Sadoulet, E. (2002). World Poverty and the Role of Agricultural Technology: Direct and Indirect Effects. *Journal of Development Studies, 38*(4), 1–26.

Diagne, A., & Demont, M. (2007). Taking a New Look at Empirical Models of Adoption: Average Treatment Effect Estimation of Adoption Rate and Its Determinants. *Agricultural Economics, 37*(3), 201–210.

Dimara, E., & Skuras, D. (2003). Adoption of Agricultural Innovations as a Two-Stage Partial Observability Process. *Agricultural Economics, 28*(3), 187–196.

Doss, C. R., Mwangi, W., Verkuijl, H., & De Groote, H. (2003). *Adoption of Maize and Wheat Technologies in Eastern Africa: A Synthesis of the Findings of 22 Case Studies*. CIMMYT Economics Working Paper 3–7, Mexico, D.F, CIMMYT.

Duflo, E., Kremer, M., & Robinson, J. (2008). How High Are Rates of Return to Fertilizer? Evidence from Field Experiments in Kenya. *American Economic Review, 98*(2), 482–488.

Fadare, A. O., Akerele, D., & Toritseju, B. (2014). Factors Affecting Adoption Decisions of Maize Farmers in Nigeria. *International Journal of Food and Agricultural Economics, 2*(3), 45–54.

Feder, G., Just, E. R., & Zilberman, D. (1985). Adoption of Agricultural Innovations in Developing Countries: A Survey. *Economic Development and Cultural Change, 33*(2), 255–298.

Foster, A. D., & Rosenzweig, M. R. (2010). *Microeconomics of Technology Adoption*. Economic Growth Centre Discussion Paper No 984, Yale University, New York.

Gollin, D., Morris, M., & Byerlee, D. (2005). Technology Adoption in Intensive Post-Green Revolution Systems. *American Journal of Agricultural Economics, 87*(5), 1310–1313.

Hailu, K. B., Abrha, K. B., & Weldegiorgis, A. K. (2014). Adoption and Impact of Agricultural Technologies on Farm Income: Evidence from Southern Tigray, Northern Ethiopia. *International Journal of Food and Agricultural Economics, 2*(4), 91–106.

Harper, J. K., Rister, M. E., Mjelde, J. W., Drees, B. M., & Way, M. O. (1990). Factors Affecting Adoption in Insect Management Technology. *American Journal of Agricultural Economics, 72*(4), 997–1005.

Kassie, M., Shireraw, B., & Muricho, G. (2011). Agricultural Technology, Crop Income and Poverty Alleviation in Uganda. *World Development, 39*(10), 1784–1795.

Kijima, Y., Otsuka, K., & Sserunkuuma, D. (2008). Assessing the Impact of NERICA on Income and Poverty in Central and Western Uganda. *Agricultural Economics, 38*(3), 327–337.

Lambkin, M., & Day, G. S. (1989). Evolutionary Processes in Competitive Markets: Beyond the Product Life Cycle. *Journal of Marketing, 53*(3), 4–20.

Langat, B. K., Ngéno, V. K., Nyangweso, P. M., Mutwol, M. J., Kipsat, M. J., Gohole, L., & Yaninek, S. (2013). *Drivers of Technology Adoption in a*

Subsistence Economy: The Case of Tissue Culture Bananas in Western Kenya. A Paper Presented at the 4th International Conference of the African Association of Agricultural Economists, September 22–25, 2013, Hammamet, Tunisia.

Liu, Y., & Wang, X. (2005). Technological Progress and Chinese Agricultural Growth in the 1990s. *China Economic Review, 16*(4), 419–440.

Matsumoto, T., Yamano, T., & Sserunkuuma, D. (2013). *Technology Adoption and Dissemination in Agriculture: Evidence from Sequential Intervention in Maize Production in Uganda*. GRIPS Discussion Paper 13–14, National Graduate Institute for Policy Studies, Tokyo.

Mendola, M. (2007). Agricultural Technology Adoption and Poverty Reduction: A Propensity-Score Matching Analysis for Rural Bangladesh. *Food Policy, 32*(3), 372–393.

Midgley, D. F. (1977). *Innovation and New Product Marketing*. New York, NY: Halstead Press, John Wiley and Sons.

Mkandawire, R., & Matlosa, K. (1994). *Food Policy and Agriculture in Southern Africa*. Harare: Southern African Research and Documentation Centre, SAPES Trust.

Muzari, W., Gatsi, W., & Muvhunzi, S. (2012). The Impacts of Technology Adoption on Smallholder Agricultural Productivity in Sub-Saharan Africa: A Review. *Journal of Sustainable Development, 5*(8), 69–77.

Oster, E., & Thornton, R. (2012). Determinants of Technology Adoption: Peer Effects in Menstrual Cup Take-up. *Journal of the European Economic Association, 10*(6), 1263–1293.

Pattanayak, S. K., Mercer, D. E., Sills, E., & Yang, J. C. (2003). Taking Stock of Agroforestry Adoption Studies. *Agroforestry Systems, 57*(3), 173–186.

Pingali, P., Bigot, Y., & Binswanger, H. P. (1987). *Agricultural Mechanization and the Evolution of Farming Systems in Sub-Saharan Africa*. Baltimore: World Bank, John Hopkins University Press.

Ravallion, M., & Chen, S. (2004). How Have the World's Poorest Fared Since the 1980s? *World Research Observer, 19*(2), 141–170.

Reimer, J. J., & Fisher, M. (2014). Are Modern Varieties Always Better? An Economic Analysis of Maize Varietal Selection. *African Journal of Agricultural and Resource Economics, 9*(4), 270–285.

Rogers, M. E. (1962). *Diffusion of Innovations: Third Edition*. New York: Macmillan Publishing Co Inc.

Simtowe, F., Kassie, M., Diagne, A., Silim, S., Muange, E., Asfaw, S., & Shiferaw, B. (2011). Determinants of Agricultural Technology Adoption: The Case of Improved Pigeon Pea Varieties in Tanzania. *Quarterly Journal of International Agriculture, 50*(4), 325–345.

Smale, M., Heisey, P. W., & Leathers, H. D. (1995). Maize of the Ancestors and Modern Varieties: The Microeconomics of High-Yielding Variety Adoption in Malawi. *Economic Development and Cultural Change, 43*(2), 351–368.

Sunding, D., & Zilberman, D. (2000). The Agricultural Innovation Process: Research and Technology Adoption in a Changing Agricultural Sector. *Handbook of Agricultural Economics, 1*, 207–261.

Suri, T. (2011). Selection and Comparative Advantage in Technology Adoption. *Econometrica, 79*(1), 159–209.

Uaiene, R. N., Arndt, C., & Masters, W. A. (2009). *Determinants of Agricultural Technology Adoption in Mozambique*. National Directorate of Studies and Policy Analysis Discussion Paper No 67E, Ministry of Planning and Development, Mozambique.

Wooldridge, J. M. (2002). *Econometric Analysis of Cross Section and Panel Data*. Cambridge, MA: The MIT Press.

World Bank. (2008). *Agriculture for Development: Overview*. World Development Report, Washington, DC. Retrieved from www.worldbank.org/INTWDR2008.

Xu, X., & Jeffrey, S. R. (1998). Efficiency and Technical Progress in Traditional and Modern Agriculture: Evidence from Rice Production in China. *Agricultural Economics, 18*(1998), 157–167.

Zivanemoyo, J., & Mukarati, J. (2013). Determinants of Choice of Crop Variety as Climate Change Adaptation Option in Arid Regions of Zimbabwe. *Russian Journal of Agricultural and Socio-Economic Sciences, 15*(3), 54–62.

4

Input Utilization and Agricultural Labor Productivity: A Gender Analysis

A. M. Rufai, K. K. Salman, and M. B. Salawu

4.1 Introduction

Agriculture employs about 65% of Africa's labor force (World Bank 2013) and the sector has been identified as the major source of income of most rural households. Sub-Saharan Africa ranks high in the world in terms of the proportion of people living in poverty, and agriculture has been identified to have the potentials of reducing poverty and promoting economic development in the region. David (2010) however explained that there is the need to improve the productivity of the sector for it to have higher impacts on aggregate economic indicators and ultimately reduce poverty. Failure to develop the agricultural sector in the region could be associated with the low performance of labor. McCullough (2015) revealed that despite the fact that countries in Sub-Saharan Africa have the highest level of value added through the agricultural sector, the region has the lowest labor productivity.

A. M. Rufai (✉) • K. K. Salman • M. B. Salawu
University of Ibadan, Ibadan, Nigeria

A. Shimeles et al. (eds.), *Building a Resilient and Sustainable Agriculture in Sub-Saharan Africa*, https://doi.org/10.1007/978-3-319-76222-7_4

In addition to land and capital, labor is identified as one of the most important and effective factors of agricultural production (Biniaz 2014). According to Oluyole et al. (2007), the availability of labor determines the quantity and quality of output as it influences planting precision, weed control, timely harvest and crop processing. Agriculture in Nigeria is labor intensive as the sector employs more than half of the labor force in the country. However, despite the high level of human resource, the contribution of the sector to the economic growth has continued to reduce over the years (Manyong et al. 2005; Mohammed-Lawal and Atte 2006). The low productivity has been associated with the fact that the sector is mostly made up of small scale farmers who still use rudimentary production techniques which makes them highly dependent on manual labor (Oluyole et al. 2013).

Family labor is an important source of manpower in agriculture. Women especially in the rural areas are known to play crucial roles in household farming activities, thus contributing significantly to the amount of labor available for agriculture. The increasing out-migration of men from rural areas and their participation in off-farm work has left agriculture more in the hands of women (Lastarria-corhiel 2006).

Women's work in agriculture has become more visible as their involvement in agricultural production has deepened in response to the economic opportunities in commercial agriculture and the rising need for them to provide for the household (Lastarria-corhiel 2006). However, despite their increased involvement in agriculture, significant differences have been identified in the level of productivity of men and women. The traditional system of division of labor where women are expected to care for the house and still participate in agricultural activities may restrict their availability for agricultural production, thus reducing the total area under crop cultivation due to labor shortages (Kwaramba 1997). Edet et al. (2016) explain that the availability of labor for farm activities determines the extent of work that can be done and ultimately the productivity of the sector.

The productivity of labor in agriculture is highly dependent on the availability of inputs and the quality of work done by labor. Okoye et al. (2008) explained that the proper allocation of inputs could assist farmers to make efficient and effective use of labor and ultimately improve productivity. The lower productivity in female-managed farms is an indication of the differences in the factors of agricultural production (i.e. both input and management) between genders. Generally, the production capacity of farm households in Nigeria is limited by their poor access to inputs such as land (as a result of the expanding population), new technology and credit facilities. However, women face greater vulnerabilities in agriculture mainly because of their poor access to inputs coupled with their relative lack of education and heavy burden of unpaid domestic work (Phillip et al. 2009). Ogunlela and Mukhtar (2009) explain that high levels of gender imbalance combined with social, religious, psychological and biological factors promote discriminations against women in terms of employment, education and access to resources.

The variations in the extent of access to inputs between gender in Nigeria and the low contribution of the agricultural sector to the GDP in the country despite the high level of labor participation makes it important to examine the extent of input utilization and understand how it influences the productivity of labor in agriculture based on gender. Such assessments are germane for the formulation of effective policies that would promote efficient use of labor and other inputs, reduce poverty in agricultural households and increase the contribution of the agricultural sector to economic growth.

The main objective of the chapter is to explore the influences of input utilization on labor productivity among men and women in Nigeria. The specific objectives of the study are to:

- assess the extent of input utilization based on gender;
- examine the extent and type of labor utilized and the productivity of labor;
- identify the factors that determine the utilization of inputs among farmers; and
- analyze the effects of input utilization on the productivity of labor.

While the efficient use of resources in agriculture is important, an effective use of the vast human resources available for agriculture in Nigeria could promote rapid agricultural development as according to

(Polyzos and Arabatzis 2005) it remains the most important factor in the production process. For agriculture to result in economic growth and reduce poverty, it is necessary for output to grow at a faster rate than the labor force in the population (Okoye et al. 2008). In Nigeria, studies such as, Anyaegbunam et al. (2010) and Okoye et al. (2008) have examined the effects of inputs on the productivity of labor while others such as Ogunniyi et al. (2012) and Umar et al. (2010) considered the differences in the productivity between males and females in their assessments. However, most of the studies on input utilization and labor productivity in Nigeria were focused on specific crops and farmers from certain parts of the country and not the country as a whole and hence lacked national representation. The availability of such information would be a substantial input in understanding the agricultural labor market dynamics in Nigeria. Also most of the studies often do not control for plot, household and village/community characteristics that could influence the gender gap in labor productivity. Ragasa et al. (2012) explains that the non-inclusion of such variables in the assessment of productivity in agriculture often leads to biased results. Also Clark (2013) opined that the inclusion of such variables does not only explain the gender gap in productivity, they also provide important insights into key variables that drive differences. This study also uses the quantile regression to assess the effects of inputs on labor productivity compared to the ordinary least squares (OLS) used in most studies. This method compared to the OLS gives a broader description of a dependent variable as a conditional function of a set of covariates (Kaditi and Nitsi 2010). Kaditi and Nitsi (2009) explain that the heterogeneity in farm data which leads to the problem of heteroskedasticity and resultant biased estimates makes quantile regression particularly suitable for its analysis.

Policies addressing agricultural productivity are masculine in nature and they often do not promote women empowerment by reducing the gender productivity differential and ensuring access to resources among women in agriculture. For the agricultural sector to promote significant economic development in Nigeria, the important roles of women, the increasing feminization of agriculture and the challenges faced by women need to be well understood and operationalized in policymaking across the country. Understanding how input utilization influences labor productivity across gender is important in making decisions about the implementation of agricultural policies and interventions in various parts of the country. The findings from such studies are also vital in the formulation of policies that are concerned with promoting food security in the country as increasing farmers productivity translates to increasing the amount of food available within the country.

4.2 Methodology

Scope of study: Nigeria is located in West Africa and shares borders with the Republic of Benin in the west, Chad and Cameroon in the east and Niger in the north. The country has a land area of approximately 923,768 square kilometers with 1.4% covered by water. It has a population of about 184,551,471 (Worldometer n.d.) and a rural population of about 93,589,090. About 90% of the rural population is employed in agriculture according to International Fund for Agricultural Development (IFAD). The three largest and most influential ethnic groups in Nigeria are the Hausa, Igbo and Yoruba, and based on this, the country can be roughly split into three regions.

Data: The general household survey (GHS) data for Nigeria collected by the National Bureau of statistics in collaboration with the Federal Ministry of Agriculture and Rural Development and the World Bank would be used for this study. The sampling frame for the data is based on the 2006 housing and population census conducted by the National Population Commission (NPopC). The sampling frame was made up of about 662,000 enumeration areas (EAs). The master frame was also generated at the local government areas (LGAs). The National Integrated Survey of Households 2007/2012 Master Frame Sample (NISH—MSF) was constructed by pooling LGAs in the master areas by state. A systematic sample of 200 EAs was then selected with equal probability across all LGAs within the state. The sample EAs for the GHS was based on a sub-sample NISH—MSF which are replicates generated from the NISH—MSF frame. A two stage sampling procedure was used in collecting the GHS data. In the first stage, the EAs (or primary sampling units) were selected based on the probability proportional to size (PPS) of the total EAs in each state and the Federal Capital Territory (FCT) and the total number of households listed in each of the EAs. The second stage involved the selection of ten households per EA by the systematic random sampling procedure.

The data is nationally representative and contains information about household characteristics, literacy rates, off-farm income generating activities, paid and unpaid employments, agricultural practices and output, labor, wage rates and farm characteristics collected from a sample of 5000 households. The data is also representative at rural and urban levels, and across the geopolitical zones of the country. The data contains adequate information that would allow the researcher answer the key questions of this research. The data can be downloaded at http://microdata.worldbank.org/index.php/catalog/1952.

4.3 Data Analysis

Descriptive statistics: This involved the use of tables, frequency distribution, means, percentages and standard deviation in analyzing the use of inputs and labor productivity by gender. The use of inputs was assessed across the six geopolitical zones in Nigeria.

Principal Components Analysis (PCA): The input index was generated using the Principal Component analysis (PCA). The PCA finds the axis system defined by principal directions of variance (α) in a given data set. It linearly transforms data into a substantially smaller set of uncorrelated variables called principal components that contain most of the information in the original data. The principal components are found by calculating the eigen vectors and eigen values of the covariance matrix. Following Kolenikov and Angeles (2004), if a random variable X with dimensions n with finite nxn variance-covariance matrix;

$$V[X] = \varepsilon \tag{4.1}$$

The principal component $(Yj)_0$ of variable $X1$ … Xn are linear combinations α_{ij} … α_{in} such that:

$$y_j = a_j x \tag{4.2}$$

where $j = 1$ … n.

Solving the eigen problem for matrix ε involves finding λ and α such that $\varepsilon\alpha = \lambda\alpha$ which gives the set of principal components weights α (or factor loadings), the linear combinations $\alpha'x$ (or factor scores) and eigen values $\lambda_1 \geq \lambda_2 \geq \ldots. \lambda n$. The variance $v[\alpha'x = \lambda_k]$ so that eigen values are the variances of the linear combinations. The index is generated as a weighted average of the variable scores with weights equal to the loadings of the first principal component.

$$c_i = \sum_{i=1}^{n} w_1 x_1 \tag{4.3}$$

where c = composite index, w = weight attributed and n = number of variables

Labor productivity: This is the ratio of output (y) and labor input (l). It is a partial productivity measure which is largely dependent on the effective use of other inputs (Organization of Economic Co-operative and Development—OECD 2011). Labor productivity (y_p) is expressed as:

$$y_p = \frac{\text{volume measure of output}}{\text{Measure of labour input}} = \frac{Y}{L} \tag{4.4}$$

The factors that explain the productivity are thereafter unveiled in a labor productivity model given as:

$$y_p = f(P, I, H) \tag{4.5}$$

where y_p = Labor productivity, P = Plot characteristics (Land size, cropping system), I = other inputs (fertilizer, herbicides, pesticides, machinery and animal traction) and H = characteristics of plot owner (socioeconomic and other household characteristics).

Equation (4.2) can be modeled explicitly as:

$$y_{fn} = \beta_0 + \beta_1 p_{fn} + \beta_2 I_{fn} + \beta_3 H_{fn} + e_{fn} \tag{4.6}$$

where y_{fn} = labor productivity on plot f in household n, p_{fn} = plot characteristics of plot f, I_{fn} = use of production inputs on plot of f, H_{fn} = characteristics of owner of plot f and e_{fn} is the error term.

Quantile Model: To estimate the labor productivity model, a quantile regression was employed. The conditional τ^{th} quantile of $yp\,(r \in [0,1])$ given a covariate vector x is expressed linearly in logarithms given a conditional quantile function

$$Q_{lny}\left(\tau / x\right) = \beta\left(\tau\right) \ln x \tag{4.7}$$

Adapting the labor productivity model in Eqs. 3 and 4 gives:

$$Q_{lny}\left(\frac{\tau}{x}\right) = \beta_0 + \beta_1\left(\tau\right) p_{fn} + \beta_2\left(\tau\right) I_{fn} + \beta_3\left(\tau\right) H_{fn} + F'\left(\tau\right) \tag{4.8}$$

To further address the problem of heteroskedasticity, the bootstrapped quantile regression was used to obtain robust standard errors. According to Singh and Xie, bootstrapping involves the use of a data sample to create a large number of samples through resampling. It is a statistical function of the form

$$T = \left(\tilde{\theta} - \theta\right) / \mathrm{SE} \tag{4.9}$$

where θ = Population parameter, $\tilde{\theta}$ = sample parameter (bootstrap) and SE = sample estimate of the standard error of $\tilde{\theta}$ which brings extra accuracy.

4.4 Results

4.4.1 Input Utilization Based on Gender

The level of input utilization on male- and female-managed plots was presented in Table 4.1. Generally less than 45% of the farmers had used fertilizer. Almost 45% of the males had used fertilizer while less than a quarter of the females had used the input on their plots. Even though females had a higher minimum quantity of 100 kg, males had a higher maximum quantity of 43,000 kg of fertilizer compared to the maximum of 20,000 kg among females. In assessing the constraints encountered by women in Agriculture in Nigeria, Fabiyi et al. (2007) explained that women identified that the high costs of inputs and late delivery of inputs, especially fertilizer, was a major constraint. Ajani and Igbokwe (2011) also identified that major constraints encountered by women in performing new roles with the feminization of agriculture were the lack of farm inputs such as fertilizer and herbicides.

For pesticides, less than 10% of the farmers had used the input on their plots. Even though a higher proportion of females (11.59%) had used the input, males used more pesticides than females as males had a maximum quantity of 40,000 kg compared to the maximum quantity of 2400 kg among female-managed plots. Similarly, over 30% of the females had used herbicides on their plots while less than a quarter of the males had used the input. However, males had a higher quantity of 10,000 kg compared to 2400 kg on female plots. For machinery/equipment, less than a quarter of males and females had used the input. Males had a higher maximum value of 40 machines/equipment while a higher proportion of women (20.86%) had used the input on their farms. This implies that even though higher proportions of females had used pesticides, herbicides and machines/equipment

Table 4.1 Use of inputs on plots managed by male and female farmers

Inputs	Yes		No		Mean	Std. dev.	Min	Max	T-test
	Freq.	%	Freq.	%					
Fertilizer (kg)									
Total	1067	41.17	1525	58.83	8594.79	26,022.87	50.00	430,000	
Male	1001	43.71	1289	56.29	9491.82	27,177.53	50.00	430,000	8.34***
Female	66	21.85	236	78.15	1792.42	12,639.14	100.00	200,000	
Pearson chi2(2) = 52.63* **									
Pesticide (kg)									
Total	217	8.37	2375	91.63	627.99	2877.45	1.00	40,000	
Male	182	7.95	2108	92.05	670.28	3139.57	1.00	40,000	1.05**
Female	35	11.59	267	88.41	411.69	456.58	4.00	2400	
Pearson chi2(2) = 4.61 **									
Herbicide (kg)									
Total	608	23.46	1984	76.54	379.02	551.84	0.02	10,000	
Male	511	22.31	1779	77.69	391.82	579.19	0.02	10,000	2.70***
Female	97	32.12	205	67.88	286.13	267.32	1.00	1200	
Pearson chi2(2) = 14.29* **									
Machinery/equipment (Pieces)									
Total	412	15.90	2180	84.10	20.88	8.58	2.00	40	
Male	349	15.24	1941	84.76	21.11	8.50	2.00	40	1.78*
Female	63	20.86	239	79.14	19.57	10.52	2.00	38	
Pearson chi2(2) = 6.31* **									
Animal traction									
Total	493	19.02	2099	80.98	3.25	3.24	1.00	40	
Male	419	18.30	1871	81.70	3.29	2.91	1.00	19	0.66
Female	74	24.50	228	75.50	2.96	4.75	1.00	40	
Pearson chi2(2) = 6.67* **									

Source: Authors' computation from GHS 2012/2013 survey data

*, **, *** show significance at 1%, 5%, and 1%, respectively

on their plots, the quantities used among them were smaller when compared to the quantities utilized on male-managed plots. The results of the T-test show there was significant difference in the quantity of fertilizer, pesticides, herbicides, machinery and equipment used on male- and female-managed farms. Ogunniyi et al. (2012), in assessing input utilization among male and female Cocoa farmers, revealed that male farmers used higher quantities of pesticides and insecticides on their farms than female farmers.

Only about 20% of the farmers had utilized animal traction on their farms. It seemed to be utilized more on female-managed plots as 24.50% of females had used the input compared to 18.30% of male-managed plots while females had also used a higher maximum number of animals (40 animals) compared to males (19 animals). However, there was no significant difference in the number of animals used for traction on both male and female farms. The results of the Pearson's chi-square statistic also revealed that there was significant difference in the use of inputs across male- and female-managed farms in Nigeria. This implies that while a higher proportion of male-managed plots used fertilizer, a higher proportion of female-managed farms used pesticides, herbicides, machinery and equipment and animal traction.

Across the zones as shown in Table 4.2, all the females in the North-West had used fertilizer on their plots while none of the females in the South-West had used the input on their farms. In the North-Central and North-East more males had used fertilizer on their plots while in the South-East and South-South there was no significant difference in the proportion of males and females that had used fertilizer on their plots. For pesticides, none of the women in the North-Central and North-West had used the input while higher proportion of females had used the input in the North-East and South-South. None of the females in the North-Central zone had also used herbicides on their farms. However, compared to males, a higher proportion of females had used herbicides in the North-East, South-East and South-West. For machinery and equipment, none of the females in the North-East had used them while a higher proportion of females had used them in the other zones except the North-Central. Animal traction had not been used by females in the North-Central and North-East while a higher proportion of males had used them in the other zones.

Table 4.2 Use of inputs across zones

	North-Central		North-East		North-West		South-East		South-South		South-West	
	Yes	No	Yes	No	Yes	No	Yes	No	Yes	No	Yes	No
	%	%	%	%	%	%	%	%	%	%	%	%
Fertilizer												
Total	33.10	66.90	46.52	53.48	81.83	18.17	26.19	73.81	14.76	85.24	4.31	95.96
Male	34.17	65.83	47.02	52.98	81.62	18.38	26.67	73.33	14.83	85.17	4.86	95.14
Female	21.62	78.38	20.00	80.00	100.00	–	25.16	74.84	14.49	85.51	–	100.00
Pesticide												
Total	3.22	96.78	11.49	88.51	5.67	94.33	13.61	86.39	11.45	88.55	1.91	98.09
Male	3.52	96.48	11.13	88.87	5.73	94.27	13.94	86.06	10.27	89.73	1.62	98.38
Female	–	100.00	30.00	70.00	–	100.00	12.90	87.10	15.94	84.06	4.17	95.83
Herbicide												
Total	12.41	87.59	16.01	83.99	22.83	77.17	45.36	54.64	30.12	69.88	5.74	94.26
Male	13.57	86.43	15.93	84.07	22.93	77.07	44.24	55.76	31.18	68.82	5.41	94.59
Female	–	100.00	20.00	80.00	14.29	85.71	47.74	52.26	26.09	73.91	8.33	91.67
Machinery/equipment												
Total	7.36	92.64	5.46	94.45	22.83	77.17	20.41	79.95	25.30	74.70	14.83	85.17
Male	7.79	92.21	5.57	94.43	22.60	77.40	19.70	80.30	24.33	75.67	14.05	85.95
Female	2.70	97.30	–	100.00	42.86	57.14	21.94	78.06	28.99	71.01	20.83	79.17
Animal traction												
Total	0.23	99.77	11.49	88.51	14.67	85.33	33.40	66.60	46.08	53.92	13.40	86.60
Male	0.25	99.75	11.71	88.29	14.67	85.33	36.97	63.03	46.39	53.61	14.05	85.95
Female	–	100.00	–	100.00	14.29	85.71	25.81	74.19	44.93	55.07	8.33	91.67

Source: Authors' computation from GHS 2012/2013 survey data

4.4.2 Labor Utilization and Productivity Among Farmers Based on Gender

4.4.2.1 Family and Hired Labor Use

The quantity of family and hired labor (men, women and children) used by male and female farmers was assessed as shown in Table 4.3. The average man-hours used of family labor on male and female plots were 199,468.70 man-hours and 175,203.60 man-hours, respectively. For hired labor, the average man-hours used on male and female plots was 588.48 man-hours and 833.32 man-hours, respectively. This indicates that, on average, female-managed plots used less family labor and more of hired when compared to male-managed ones. The average of hired

Table 4.3 Labor input use

	Mean	Std. dev.	Min	Max	T-test
Family labor					
Total	196,641.00	804,835.00	0.00	23,400,000.00	0.59
Male	199,468.70	824,542.90	0.00	23,400,000.00	
Female	175,203.60	636,546.90	8.00	6,865,218.00	
Hired labor					
Total hired labor					
Total	617.00	3130.44	0.00	87,499.38	2.64***
Male	588.48	2761.82	0.00	85,261.08	
Female	833.32	5478.61	0.00	87,499.38	
Men					
Total	394.44	2354.33	0.00	85,005.00	0.27
Male	390.17	2374.17	0.00	85,005.00	
Female	426.76	2201.47	0.00	31,368.00	
Women					
Total	198.61	1972.33	0.00	87,395.88	1.81**
Male	172.06	1032.25	0.00	30,809.63	
Female	399.93	5033.54	0.00	87,395.88	
Children					
Total	23.96	284.06	0.00	12,804.00	2.91***
Male	26.24	301.77	0.00	12,804.00	
Female	6.63	41.24	0.00	185.60	

Source: Authors' computation from GHS 2012/2013 survey data
*significant at 1%, ** significant at 5%, *** significant at 1%

men, women and children used across the plots were 394.44, 198.61 and 23.96 man-hours respectively.

Generally, Female-managed plots had a higher mean of 426.76 man-hours from hired men when compared to male-managed plots. However, male-managed plots had a higher maximum labor (85,005 man-hours) spent working on plots by hired men. Hired females[1] were utilized more on female-managed plots as they had a higher average and maximum labor of 399.93 man-hours and 87,395.88man-hours, respectively compared to 172.06 man-hours and 30,809.63 man-hours on male-managed plots. More child labor was used on male-managed plots compared to those managed by females. The average and maximum time spent by children on male plots was 26.24 man-hours and 12,804.00 man-hours. However, on female plots, an average labor of 6.63 man-hours was used by children and the maximum labor was 185.60 man-hours. There was a significant difference in the quantity of total hired labor, female hired labor and child hired labor between male- and female-managed plots in Nigeria; however, there was no significant difference in the amount of family labor and hired male labor.

Across the zones, as revealed in Table 4.4, female farmers in the South-South and South-West had higher means of time spent by male labor on their farms while in the North-Central and South-South they had a higher mean of time spent by female labor when compared to male-managed farms. The proportions of family and hired labor used on male and female plots vary with location in Nigeria as Obasi and Kanu (2014) found that male farmers had more access to hired labor than their female counterparts in Imo state while Ogunniyi et al. (2012) found that females used more hired labor and less family labor when compared to men in Ondo state.

4.4.2.2 Labor Productivity

In Table 4.5, male-managed plots had higher value of outputs with an average of ₦159,344.10 and a maximum of ₦3,059,000 while female-managed ones had an average of ₦133,138.60 and a maximum value of ₦1,590,000. The amount of labor utilized on male-managed plots was higher as they had an average of 208,266.60 man-hours while females

Table 4.4 Labor input used by zones

	North-Central		North-East		North-West		South-East		South-South		South-West	
	Mean	Std. dev.	Mean	Std. dev.	Mean	Std. dev.	Mean	Std. dev.	Mean	Std. dev.	Mean	Std. dev.
Family labor												
Total	173,945.1	746,172.1	165,874.2	576,972.0	222,716.3	649,088.9	131,989.9	409,904.9	337,047.5	1,628,264.0	174,185.5	434,148.7
Male	179,731.8	776,536.6	168,178.9	582,068.6	224,695.6	652,612.9	133,669.1	384,525.7	340,584.5	1,745,948.0	165,945.0	425,362.1
Female	111,699.9	240,839.1	45,799.8	110,686.9	55,039.0	75,804.2	128,414.9	460,616.4	323,565.8	1,077,051.0	237,706.5	502,335.8
Hired labor												
Total hired labor												
Total	723.1	4151.7	552.9	2886.4	536.88	1352.4	524.0	1882.5	799.9	4985.9	714.4	3415.8
Male	727.1	4331.9	558.0	2913.2	540.43	1359.6	546.3	1467.5	567.0	1552.8	635.7	3322.8
Female	678.9	922.1	284.8	422.2	236.39	283.1	476.5	2556.1	1687.8	10,521.8	1321.4	4093.2
Men												
Total	486.1	4098.8	384.3	2711.5	394.1	1257.1	375.7	1769.7	317.2	658.0	396.9	1643.1
Male	491.8	4280.5	388.9	2736.9	396.7	1264.1	375.4	1272.7	312.5	588.7	290.8	919.9
Female	425.2	689.9	139.7	328.8	174.0	245.9	376.3	2526.4	335.0	878.7	1211.8	4106.2
Women												
Total	220.4	648.2	147.1	939.9	118.4	427.7	131.2	643.7	464.3	4954.7	248.9	2149.5
Male	218.2	646.1	147.5	947.9	119.1	430.1	149.2	761.8	231.8	1436.9	267.4	2283.1
Female	243.8	678.5	122.4	330.0	52.6	74.7	92.9	245.9	1350.6	2051.6	106.9	244.3
Children												
Total	16.5	96.2	21.5	194.8	24.4	126.4	17.1	129.1	18.5	102.6	68.9	886.2
Male	17.1	99.9	21.5	196.5	24.6	127.1	21.7	152.3	22.7	114.8	77.5	941.8
Female	9.9	38.2	22.6	58.9	9.78	21.9	7.3	51.8	2.2	8.6	2.7	9.0

Source: Authors' computation from GHS 2012/2013 survey data

Table 4.5 Labor productivity based on gender

Productivity	Mean	Std. dev.	Min	Max
Value of output (₦)				
Total	156,290.90	226,417.70	500.00	3,059,000.00
Male	159,344.10	226,417.70	500.00	3,059,000.00
Female	133,138.60	191,071.80	500.00	1,590,000.00
Total labor (man-hour)				
Total	204,576.20	875,834.70	21.00	23,400,000.00
Male	208,266.60	902,707.60	21.00	23,400,000.00
Female	176,592.20	636,619.80	162.00	6,865,532.00
Labor productivity (₦/man-hour)				
Total	48.93	275.64	0.01	9976.19
Male	49.07	287.49	0.01	9976.19
Female	47.86	159.59	0.02	1376.81

Source: Authors' computation from GHS 2012/2013 survey data

had an average of 176,592.20 man-hours. Labor productivity was therefore higher on male-managed plots with an average of ₦49.07/man-hour and a maximum value of ₦9976.19/man-hour compared to female-managed plots with an average of ₦47.86/man-hour and a maximum of ₦1376.81/man-hour. Palacios-Lopez and Lopez (2015) explained that in Sub-Saharan Africa, land and labor productivity were higher in plots managed by male headed households; however, the gender difference was greater for labor productivity.

4.4.3 Effects of Input Utilization on Labor Productivity by Gender

4.4.3.1 Gender-Specific Labor Productivity Model (Male)

In the male model in Table 4.6, the results from the OLS when compared to the results from the quantile regression seem very similar. However, while three variables were significant in the OLS, four (4), four (4), three (2) and two (2) variables were significant in the 25th, 50th, 75th and 90th quantiles, respectively. The variation in the number of significant variables across the quantiles is an indication of the heterogeneity that exists in farm data[2]. The R2 revealed that the variables in the model

Table 4.6 Labor productivity model (male)

Explanatory variables	OLS	Quantile regression 25th quantile	50th quantile	75th quantile	90th quantile
Input use index	0.036	0.067	−0.004	−0.004	−0.114**
North (dummy)	2.608***	0.818***	0.407***	0.261***	0.179
Age	0.118	0.027	0.015	0.011	0.043
Education	−0.024	−0.001	−0.008	−0.028	0.046
Rural (dummy)	0.073	−0.241**	−0.211*	−0.298***	−0.091
Multiple cropping	0.556***	0.001***	0.001***	0.001**	0.001***
Own land	−0.363*	−0.563	−0.068	−0.195	−0.093
Land size	0.134	0.001*	0.001***	0.001	0.001
Household size	−0.257	−0.759	−0.004	−0.002	0.019
Has savings	−0.416	−0.255	0.026	0.197*	0.054
R2	0.561	–	–	–	–
Adjusted R2	0.356	–	–	–	–
Pseudo R2	–	0.1775	0.1708	0.1489	0.1266

Source: Authors' computation from GHS 2012/2013 survey data
*significant at 10% level; **5% level; ***1% level

explained about 56.1% of the variations in land productivity among males. Also the variables representing land size and savings that were not significant in the OLS were found to be significant in the 25th, 50th and 95th quantile regression.

The coefficient of the use of index was significant in the 90th quantile. The index, even though not significant, had a positive effect only in the 25th quantile while it was negative in the other quantiles. This implies that input use only had an increasing effect on labor productivity among farmers who had the lowest productivity. Labor productivity increased significantly with farmers in the northern zones in the 25th, 50th and 75th, respectively. The coefficients indicate that the labor productivity in the northern zones was highest in the lowest quantile and decreased across the quantiles. Labor productivity also reduced significantly in rural areas across all the quantiles except in the 90th quantile while it increased among land owners in the 25th and 50th quantiles respectively. Labor productivity increased among male farmers who practiced multiple cropping across all the quantiles.

Table 4.7 Labor productivity model (female)

Explanatory variables	OLS	Quantile regression 25th	50th	75th	90th
Input use index	−0.134	0.067	−0.074	−0.111	−0.152***
North (dummy)	−0.341	1.397***	0.761***	−0.137	0.127
Age	0.185	0.152	−0.076	−0.088	−0.039
Education	0.268	0.079	0.099	0.973*	0.032
Rural (dummy)	0.417	−0.005	0.286	0.959*	0.222
Multiple cropping	−0.387	0.001***	−0.002	0.001***	0.001***
Own land	2.015	−0.649	−0.500*	−0.479	2.544*
Land size	1.945	0.001	2.501*	0.001	2.232
Household size	−0.001	−0.055	−0.009	0.155*	0.146**
Has savings	0.068*	0.217	0.388*	0.147	0.347
R2	0.539	–	–	–	–
Adjusted R2	0.383	–	–	–	–
Pseudo R2	–	0.1383	0.1167	0.1202	0.1919

Source: Authors' computation from GHS 2012/2013 survey data
*significant at 10% level; **5% level; ***1% level

4.4.3.2 Gender-Specific Labor Productivity Model (Female)

In the female labor productivity model on Table 4.7, only one variable was significant in the OLS model while two were significant in the 25th quintile and four variables were significant in each of the other quantiles. The R2 reveals that the variables in the model explain about 53.9% of the variations in land productivity among females. Just like the male model, the input use index had a positive effect only in the 25th quantile while it was negative in the other quantiles. It was also only significant in the 90th quantile. Labor productivity among women in the north increased significantly in the 25th and 50th quantile while it increased significantly in the rural areas in the 75th quantile. Labor productivity increased among female farmers who practiced multiple cropping across all the quantiles while it decreased significantly among women who owned land in the 50th quantile and increased among those in the 90th quantile. The labor productivity among women also increased with household size in the 75th and 90th quantiles while it increased with women that had savings in the 50th quantile.

4.5 Conclusion

The extent of input utilization and its effect on labor productivity based on gender was examined in the study. The use of inputs was generally low among both male and female farmers. Compared to males, female-managed farms in the South-South and South-West used more of hired labor. More man-hours were utilized on male-managed plots (208,266.60 man/hours) when compared to those managed by females (176,592.20 man/hours). Males had a higher level of labor productivity (₦49.07/man-hour) when compared to females (₦47.86/man-hour). Labor productivity decreased significantly with input use among both male and female as their level of productivity increased.

4.5.1 Recommendations

- Generally, only very few farmers had utilized inputs on their farms. The generally low productivity of the agricultural sector in Nigeria can be improved extensively through the availability of agricultural inputs and its efficient use by farmers. The agricultural policy of the country should be revised and effectively implemented by the Federal Government of Nigeria. This would not only promote farmer's access to resources but would also improve productivity.
- The use of fertilizer was particularly low among female-managed plots in the southern zones. Generally, even though more of them used other inputs, the quantity utilized was lower when compared to male-managed plots. The implementation of gender-sensitive policies should be strengthened to increase the access of female farmers to production inputs.
- Females also used more hired labor in their production activities. The agricultural labor market in Nigeria needs to be standardized and productivity to improve the performance of labor and promote labor use efficiency among farmers.
- Labor productivity decreased significantly with input use among both male and female in the highest productivity quantile. The Ministry of Agriculture and Rural Development needs to organize trainings to

build the capacity of farmers so as to enhance their resource use skills and production efficiency. With the increasing participation of women in agriculture, female farmers should be encouraged to participate in such training programs.

Appendix

Table 4.8 Descriptive statistics of explanatory variables used

	Total		Male		Female	
Variables	Freq.	%	Freq.	%	Freq.	%
Age						
19–30	151	5.83	149	6.51	2	0.66
31–40	479	18.48	462	20.17	17	5.63
41–50	613	23.65	562	24.54	51	16.89
51–60	632	24.38	543	23.71	89	29.47
>60	717	27.66	574	25.07	143	47.35
Mean	52.53		51.53		60.16	
Standard dev.	14.79		14.79		12.37	
Education						
No education	1190	45.91	1034	45.15	156	51.66
Primary	596	22.99	546	23.84	50	16.56
Secondary	659	25.42	583	25.46	76	25.17
Higher	147	5.67	127	5.55	20	6.62
Number of males in household						
0	62	2.39	–	–	62	20.53
1–3	1581	61.00	1407	61.44	174	57.62
4–6	814	31.40	754	32.93	60	19.87
>6	135	5.21	129	5.63	6	1.99
Household has savings						
Yes	697	26.89	618	26.99	79	26.16
No	1672	73.11	1672	73.01	223	73.84
Own land						
Yes	70	2.70	59	2.58	11	3.64
No	2231	97.30	2231	97.42	291	96.36
Multiple cropping						
yes	1750	67.52	1529	66.77	221	73.18
No	842	32.48	761	33.23	81	26.82
Mono cropping						
Yes	738	28.47	662	28.91	76	25.17
No	1854	71.53	1682	71.09	226	74.83

Source: Authors' computation from GHS 2012/2013 survey data

Table 4.9 Variance inflation factor (VIF) test

	VIF	Tolerance	R-squared
Fertilizer	1.11	0.9004	0.0996
Pesticide	1.16	0.8642	0.1358
Herbicide	1.16	0.8646	0.1354
Animal traction	1.28	0.7789	0.2211
Mach/equip	1.21	0.8294	0.1706
Age	1.10	0.9095	0.0905
Education	1.06	0.9471	0.0529
Cropping system	1.13	0.8851	0.1149
No of males	1.03	0.9688	0.0312
No of animals	1.04	0.9591	0.0409
Savings	1.02	0.9851	0.0149
Mean VIF	1.12		

Source: Authors' computation from GHS 2012/2013 survey data

Table 4.10 Result of PCA for inputs

The eigenvalues of the PCA for the inputs show that the first component has a variance of 1.28, and the second component has a variance of 1.13. The variance of the first and second component represents 25.7% and 22.6% of the total variance in the access to healthcare. The first and second components explain 48.3% of the total variance of the five indicators in the input use index. In the first component, pesticide and animal traction had a negative and reducing effect on the healthcare index while in the second component, only fertilizer had reducing effects

Component	Eigenvalue	Difference	Proportion	Cumulative
Comp1	1.285	0.155	0.257	0.257
Comp2	1.131	0.169	0.226	0.483
Comp3	0.961	0.071	0.192	0.676
Comp4	0.890	0.158	0.178	0.853
Comp5	0.732		0.146	1.0000

Principal components (eigenvectors) for first two components

Variable	Comp1	Comp2
Fertilizer	0.353	−0.517
Pesticide	−0.104	0.618
Herbicide	0.149	0.519
Animal traction	−0.689	−0.002
Mach/equip	0.606	0.283

Notes

1. Following International Rice research institute—IRRI (1991) and Khurana R.M. 1992, the time spent by hired female and child labor was adjusted to the hired male labor by multiplying the time they spent by 0.75 and 0.50, respectively.
2. The heterogeneity in the data was confirmed with the F test which showed that significant difference existed in the productivity of labor across the quantiles. F = 75.23 ($p < 0.01$).

References

Ajani, E. N., & Igbokwe, E. M. (2011). Implication of Feminization of Agriculture on Women Farmers in Anambra State, Nigeria. *Journal of Agricultural Extension, 15*(1). https://doi.org/10.4314/jae.v15i1.4.

Anyaegbunam, H. N., Okoye, G. N., Asumugha, M. C., Ogbonna, T. U., Madu, N. N., & Ejechi, M. E. (2010). 'Labour Productivity Among Small Holder Cassava Farmers in Southeast Agroecological Zone', Nigeria. *African Journal of Agricultural Research, 5*(21), 2882–2885.

Biniaz, A. (2014). Labour Productivity and Factors Affecting Its Demand in Paddy Farms in KB Province, Iran. *TI Journal, Agricultural Science Developments, 3*(7), 251–255.

Clark, J. T. (2013). *Understanding the Gender Based Productivity Gap in Malawi's Agricultural Sector*. An unpublished thesis submitted to the Faculty of the graduate school of Arts and Sciences of Georgetown University, Washington, DC, April 16.

David, S. (2010). The Rural Non-farm Economy, Livelihood Strategies and Household Welfare. *African Journal of Agricultural Research, 4*(1), 82–109.

Edet, B. N., Edet, M. E., & Agom, D. I. (2016). Impact of Institutional Funding on Agricultural Labour Productivity in Nigeria: A Co-integration Approach. *Agricultural Science Research Journal, 6*(2), 49–55.

Fabiyi, E. F., Danladi, B. B., Akande, K. E., & Mahmood, Y. (2007). Role of Women in Agricultural Development and Their Constraints: A Case Study of Bilri Local Government Area, Gombe State, Nigeria. *Pakistan Journal of Nutrition, 6*(6), 676–680.

International Rice Research Institute (IRRI). (1991). *Basic Procedures for Agroeconomic Research*. International Rice Institute P.O.Box 933, 1099 Manila, Philippines. ISBN 97-1-22-0007-8. Retrieved from www.irri.org.

Kaditi, E. A., & Nitsi, E. I. (2009). *A Two-Stage Productivity Analysis Using Bootsrapped Malmquist Indices and Quantile Regression*. No 52845, selected paper presented at the 111th EAAE-IAAE seminar, Canterbury, UK, June 26–27.

Kaditi, E. A., & Nitsi, E. I. (2010). *Applying Regression Quantiles to Farm Efficiency Estimation*. Selected paper prepared for presentation at the Agricultural and Applied Economics Association 2010. AAEA, CAES and WAEA Joint Annual Meeting, Denver, Colorado, July 25–27.

Khurana R. M. (1992). *Agricultural Development and Employment Patterns in India: A Comparative Analysis of Punjab and Bihar*. The concept publishing company, A/15–16, Commercial block, Mohan Garden New Delhi 110059, India.

Kolenikov, S., & Angeles, G. (2004). *The Use of Discrete Data in PCA: Theory, Simulations and Applications to Socio Economic Indices*. Working Paper WP – 04 – 85. MEASURE Evaluation, Chapel Hill. Retrieved from http://www.cpc.unu.edu/measure/publications/wp-04-85.

Kwaramba P. K. (1997, December). *The Socio-Economic Impact of HIV/AIDS on Communal Agricultural Systems in Zimbabwe*. FES Economic Advisory Services Working Paper No 19.

Lastarria-corhiel, S. (2006). *Feminization of Agriculture: Trends and Driving Forces*. Background paper for the World development Report 2008. November 2006.

Manyong, V. M., Ikpi, A., Olayemi, J. K., Yusuff, S. A., Omonona, B. T., Okoruwa, V., & Idachaba, F. S. (2005). *Agriculture in Nigeria: Identifying Opportunities for Increased Commercialization and Investment*. Ibadan: IITA.

McCullough, E. B. (2015). *Labor Productivity and Employment Gaps in Sub-Saharan Africa*. Policy Research Working Paper 7234, World Bank Group, Africa Region, Office of the Chief Economist, April.

Mohammed-Lawal, A., & Atte, O. A. (2006). An Analysis of Agricultural Production in Nigeria. *African Journal of General Agriculture, 2*(1).

Obasi, O. O., & Kanu, W. N. (2014). Gender Access to Farm Labour and Coping Strategies: Implication for Food Productivity in Imo State, Nigeria. *International Journal of Development and Sustainability, 3*(8), 1777–1781.

Ogunlela, Y. I., & Mukhtar, A. A. (2009). Gender Issues in Agriculture and Rural Development in Nigeria: The Role of Women. *Humanity and Social Sciences Journal, 4*(1), 19–30.

Ogunniyi, L. T., Ajao, O. A., & Adeleke, O. A. (2012). Gender Comparison in Production and Productivity of Cocoa Farmers in Ile Oluji Local Government

Area of Ondo State, Nigeria. *Global Journal of Science Frontier Research Agriculture Biology, 12*(5), 59–63.

Okoye, B. C., Onyenweaku, C. E., Ukoha, O. O., Asumugha, G. N., & Aniedu, O. C. (2008). Determinants of Labour Productivity on Small Holder Farms in Anambra State, Nigeria. *Scientific Research and Essay, 3*, 559–561.

Oluyole, K. A., Adebiyi, S., & Adejumo, M. O. (2007). An Assessment of the Adoption of Cocoa Rehabilitation Techniques Among Cocoa Farmers in Ijebu East Local Government Area of Ogun State. *Journal of Agricultural Research and Policies, 2*(1), 56–60.

Oluyole, K. A., Dada, O. A., Oni, O. A., Adebiyi, S., & Oduwole, O. O. (2013). Farm Labour Structure and Its Determinants Among Cocoa Farmers in Nigeria. *American Journal of Rural Development, 1*(1), 1–5.

Organisation for Economic Co-Operation and Development—OECD. (2011). *Measuring Productivity: Measurement of Aggregate and Industry-Level Productivity Growth*. An OECD Manual. Retrieved from www.SourceOECD.org.

Palacios-López, A., & López, R. (2015). The Gender Gap in Agricultural Productivity: The Role of Market Imperfections. *The Journal of Development Studies, 51*(9), 1175–1192.

Phillip, D., Nkoya, E., Pender, J., & Oni, O. A. (2009). *Constraints to Increasing Agricultural Productivity in Nigeria: A Review. A Nigeria Strategy Support Program (NSSP)*. Background paper No NSSP 006, September.

Polyzos, S., & Arabatzis, G. (2005). Labor Productivity of the Agricultural Sector in Greece: Determinant Factors and Interregional Difference Analysis. *Discussion Paper Series, Department of Planning and Regional Planning, School of Engineering, University of Thessaly, 11*(2), 209–226.

Ragasa, C. Berhane, G., Tadesse, F., & Taffesse, A. S. (2012). *Gender Differentials in Access to Extension Service and Agricultural Productivity*. Ethiopia strategy program 2, ESSP Working paper 49.

Umar, H. S., Luka, E. G., & Rahman, S. A. (2010). Gender Based Analysis of Labour Productivity in Sesame Production in Doma Local Government Area of Nasarawa State, Nigeria. *Patnsuki Journal, 6*(2), 61–68.

World Bank. (2013). Agriculture in Africa. Fact Sheet: The World Bank and Agriculture in Africa.

Worldometers. (n.d.). *Nigeria Population Growth Rate*. Retrieved February 11, 2015, from www.worldometers.info.

5

Evaluation of Women's On-Farm Trial of Drought Tolerant Maize in Southern Guinea Savannah Agro-Ecological Zone of Nigeria

O. E. Ayinde, T. Abdoulaye, G. A. Olaoye, and A. O. Oloyede

5.1 Introduction

The importance of food security in Africa cannot be overemphasized. This is seen in the growing discussions and efforts made in meeting the food security challenge such as the translation from Millennium Development Goals (MDGs) to Sustainable Development Goals (SDGs). To solve the food security challenge, there is a need to increase agricultural production through technological innovations to boost production in Africa. However, this must be achieved against a backdrop of issues such as climate change and droughts amongst others (Global Food Security Index 2015). Drought is the most devastating and costly challenge to crop production because farmers in Africa practise rain-fed

O. E. Ayinde (✉) • G. A. Olaoye • A. O. Oloyede
University of Ilorin, Ilorin, Nigeria

T. Abdoulaye
International Institute of Tropical Agriculture, Ilorin, Nigeria

A. Shimeles et al. (eds.), *Building a Resilient and Sustainable Agriculture in Sub-Saharan Africa*, https://doi.org/10.1007/978-3-319-76222-7_5

agriculture (Ayinde et al. 2016). Many farmers in Africa are faced with significant reduction in yield due to drought. Fisher et al. (2015) stated that yield losses of about 10–25% are recorded from around 40% of Africa's maize-growing areas due to drought stress. It is estimated that by 2030 drought and rising temperatures could render Africa's maize-growing areas unsuitable for current varieties (CGIAR Big Facts). In addition, it is expected that by 2050 population growth and changing diet will resultantly increase the demand for maize in the developing countries double fold (CIMMYT and IITA 2015). There is also a prediction of an annual 1.3% growth rate in demand for human consumption of maize in the developing world until 2020 (Ortiz et al. 2010).

This has implications to sub-Saharan Africa, where an array of factors contributes to a sharply increasing demand for maize, including maize being a staple food for an estimated 50% of the population (Olaniyan 2015) and its importance in addressing the issues related to food security and economic wellbeing. Maize is the most widely grown crop and in terms of food security, it is the most important cereal crop in sub-Saharan Africa (Olaniyan 2015, IITA). Out of 53 countries in sub-Saharan Africa, 46 grow maize. For instance, Nigeria has the largest land area under maize production (seventh in the world and 2.4% of the total land area) and ranks amongst the top producers (FAOSTAT 2010). Bamire et al. (2010) indicated that maize production in Nigeria is of strategic importance for food security and the socio-economic stability of the country and sub-regions in sub-Saharan Africa. However, recurring droughts has been a continuous challenge to the production of this important crop by drastically reducing yields and livelihoods. Report has it that around 25% of the maize crop suffers frequent drought, with losses of up to half the harvest in the country (CIMMYT 2013). As a result, new maize varieties will have to be developed quickly and growing in farmers' fields in the next few years if we are to avoid widespread famine in Africa (CGIAR Big Facts).

In a response to these challenges, the Drought Tolerant Maize for Africa (DTMA) project has made releases of 160 DT maize varieties, between 2007 and 2013 (Fisher et al. 2015). The International Institute of Tropical Agriculture (IITA) and International Maize and Wheat Improvement Center (CIMMYT) have been the leading force in DT maize variety research in Nigeria (Ayinde et al. 2016). The DT maize varieties, which have demonstrated superiority for grain yield, have been selected annually for testing under farmer's growing conditions in the northern and southern guinea savanna zones of the country. Ayinde et al. (2013) stated that the use of DT maize varieties stabilize maize yields in drought prone ecologies in the country. The development, dissemination and adoption of DT maize varieties, therefore, have the potential of reducing vulnerability and food insecurity (Bamire et al. 2010).

Central to these are the small-scale, resource-poor farming households living in the more marginal rain-fed agricultural areas. In Nigeria, agriculture is predominantly in the hands of rural smallholder farmers who are responsible for more than 70% of the agricultural production in the country (Enete and Amusa 2010). According to African Development Bank (AFDB) (2015), women constitute almost 50% of the agricultural labour force, yet they receive a significantly lower share of income with an estimated rural wage gap of 15–60% between men and women in the same sub-sector.

Despite women's contribution to agriculture, the rural economy and food security in the country, their ability to obtain agricultural inputs is directly constrained by gender discrimination (Simonyan et al. 2011). Soyemi (2014) stated that in a bid to increase productivity of rural farmers in the country, agricultural policies and programmes have focused on development and transfer of appropriate technologies. However, the constraint to such an approach especially for women farmers are that most agricultural technologies are being designed on the assumption that farm managers are men (Simonyan et al. 2011). In addition, even when women own the land, they tend to have limited access to financing, quality inputs, extension services and knowledge of agricultural practices. Ajadi et al. (2015), Soyemi (2014), Adeyemo et al. (2015), and Koyenikan and Ikharea (2014) found that women had less access to agricultural resources and information on agricultural technologies. Beuchelt and Badstue (2013) in a study emphasized that the reduction of gender disparities and the empowerment of women leads to better food and nutrition security for households and significantly strengthens other development outcomes such as child education.

It is therefore necessary for farmers to have access especially to agricultural technologies as this will contribute to both food security and economic growth. One of the ways to creating access to agricultural technologies and innovations is the on-farm trial method. On-farm trial is a farming systems approach which includes farmers' participation and helps farmers gain access to new technology earlier, improves farmers' knowledge and experience and encourages feedback between farmers and researchers that can allow for modifications as the growing season progresses. The method also improves farmers' capacity and expertise for conducting collaborative research and can encourage wider adoption of the technology. However, there has been lesser participation of women in on-farm trials. Women's involvement in on-farm trials has been limited and they have been left out of decision-making and evaluation of trials as well as related training (Lahai 1994). In a study in Sierra Leone, Lahai (1994) found that the focus of the extension instructor for the on-farm trials was on the women's husbands despite the fact that the women contributed significantly to the labour for the on-farm research. Saito et al. (1994) also noted that new technology may not be adopted because of the failure to adequately involve women in technology design and implementation. This study is therefore designed to evaluate women's on-farm trial of DT maize varieties in Southern Guinea Savannah zone of Nigeria. According to Norman et al. (1995), evaluation of on-farm trials includes the technological, economic and social analysis. Based on this, the specific objectives of the study are to identify women farmers' preference and evaluate the profitability of the women farmers in new variety production. Women's integration in the testing of agricultural technological innovation such as DT maize has to be examined to enhance gender equity in the use of agricultural technology (DT maize) so as to improve adoption and productivity. Consequently, a gender-balanced agricultural growth is critical to increases in food security and attainment of the SDGs (Ayinde et al. 2012).

5.2 Materials and Methodology

5.2.1 Description of Experimental Materials

Ten DT open pollinated varieties (OPVs) of maize were tested in farmers' fields in Kwara, Niger and Oyo states of the southern guinea savannah

(SGS), using on-farm trial approach through women groups. A total of ten kilograms of seeds were distributed among the farmers. The seeds of the improved DT OPVs were provided by IITA to the participating farmers. Furthermore, apart from the on-farm trial, approximately five kilograms of seeds were distributed to women as seed drop to promote production of DT maize varieties.

5.3 Method

Seven villages were selected for study. Four villages out of the selected villages had participated in the DT on-farm trial before while the remaining three villages were new. Sensitization was done through visits to the villages. Several meetings were held with the women in the selected villages and collaborators (scientists) from Agricultural Development Projects (ADP) extension agents were in attendance. The purpose for this was proper briefing on the objectives, scope and modality for executing the activities of the project. Women groups were formed in each of these villages: Mokwa, Kishi, Omupo, Lajiki, Arandun, Ballah and Isanlu Isin. This makes a total of seven groups. Each group consists of a minimum of 10 members and a maximum of 20 members. This was done for effectiveness in communication within the groups.

The women groups were trained on correct agronomic practice to enable them understand how to carry out the on-farm trials. The advantages of DTMA were highlighted to the participating women farmers. A plot was marked out for each group for on-farm trials. The plot size for the on-farm trials which included two DT maize varieties and a farmer's variety (sandwiched between the two DT varieties) as check was 10 m × 10 m. Maize was planted at an intra-row spacing of 50 cm on 75 cm wide ridges. The fields were kept weed free by the farmers while NPK Fertilizer was applied as split dosage at 4 and 7 Weeks After Planting (WAP) respectively using the recommended dosage in each zone.

The women groups were given extra seeds to plant on their individual farms. Seeds were also given to women who were not part of the group but who were interested at some of the selected villages. Two monitoring tours were undertaken to each farming community during the year. These were at the vegetative and flowering phases of the crops, respectively. The visit at the vegetative phase was to further explain the details of the protocols to ensure compliance to the details in the protocols and to collect data at this growth stage. The visit at the flowering phase was essentially for data collection at this phase. Collection of crop performance data from the on-farm trials was undertaken in conjunction with scientists and participating farmers especially during the flowering period. Data were collected essentially on yield parameters. Two estimates of yield (i.e. based on cob and seed yields) were obtained. Ears per plant (EPP) was estimated as a proportion of number of cobs harvested to total harvestable plants while seed to cob yield ratio was expressed as a proportion of yield estimated from cob and seed yield, respectively. Other socio-economic data were collected through questionnaires.

The sampling technique consists of a two-stage stratified sampling. At the first stage, eight women farmer groups were selected. The second stage involved a random selection of ten women farmers per group per location (village). The total number of farmers selected was 80.

5.3.1 Analytical Technique

Data collected from the trials were analysed. The analytical techniques used include descriptive statistic, ranking method and farm budgeting tool. Descriptive statistic such as frequency, averages, mode, mean and ranking technique which involved the use of a 3-Likert scale to analyse the socio-economic characteristics of the women farmers and their preference scores in the study area. Farm budgeting analysis was used to analyse the profitability of the DT maize varieties in the study area.

5.3.2 Farm Budgeting Analysis

The Gross margin analysis was used. Input quantities, factor prices, physical output and gross returns were used. Since the fixed cost constitutes a negligible portion of the total costs of production, the gross margin analysis was employed and used. It is given as:

$$GM_j = TR_j - TVC_j \tag{5.1}$$

where GM_j = Gross Margin (₦/ha), TR_j = Total Revenue (₦/ha), TVC_j = Total Variable Cost (₦/ha)

Returns to investment (RI) were also obtained. This was done by dividing gross margin (GM) by the total variable cost of production per hectare. The implicit form is shown below:

$$GM_j = TR_j - TVC_j \times \frac{100}{1} \tag{5.2}$$

where TR and TVC are as defined earlier.

5.4 Results and Discussion

5.4.1 Socio-Economic Characteristics of Women Farmers

The socio-economic profile of the women farmers is presented in Annex (Table 5.1). The result of the analysis revealed that the average age of farmers in the study area is 43 years with the oldest being 65 years and the youngest 22 years old. This implies that the women farmers are experienced in farming activities. About 23% of the female farmers were not educated. All the women farmers were married. The female farmers had an average farm size of 1.62 hectares. This implies that the women are less privileged to inputs and less involved in the decision-making process.

5.4.2 Women Farmer Variety Preference

Tables 5.2, 5.3, 5.4, 5.5, 5.6, 5.7, and 5.8 in Annex revealed the women farmers preference per location. The women farmers at Lajiki preferred the TZEEI 95 × TZZEE 58 variety than the other varieties. This could be a result of the fast growth and deep green leaf colour of the plant.

The women farmers at Ballah preferred the TZEEI 81 × TZZEE 95 variety than the other varieties. This could be a result of the attractive look and high vigour of the stands. The result for Arandun showed that the women preferred the EVDT 99 variety the most. The variety was preferred because it has a lot of grain although the cob is small with small seeds than the other varieties and also because EVDT 99 was fast in growth and was able, to some extent, better withstand the climate than the other varieties. The women also indicated that had there been more rain, the variety would have produced better yield. The women group at Omupo and Isanlu Isin preferred the IWDC3 SYN × 21 White DT STR SYN-DT C1 variety than the others. This is due to the fact that the variety had fast growth and was tolerant to drought. The women group at Mokwa preferred the 2013 TZE-W DT STR. This is because the variety had bigger cobs.

5.4.3 Profitability of On-Farm Trial

Table 5.9 in Annex shows the profitability of the on-farm trials. The result shows that, at Lajiki, the TZEEI 95 × TZZEE 58 variety gave the highest yield (1560 kg/ha), highest profit of ₦33,400 and highest returns to investment of 90.8%. This implies that for every 1 naira spent, 90 kobo (0.90) was gained as returns. At Ballah, the TZEEI 81 × TZZEE 95 variety produced the highest yield (1539 kg/ha), highest profit of ₦32,985 and the highest (90.9%) returns to investment. This implies that the farmers gained 91 kobo (0.91) for every 1 naira spent. At Arandun, the 99TZEE-Y-STR variety produced the highest yield of 2649 kg/ha, the highest profits of ₦81,005 and the highest returns to investment of 212%. This implies that for every 1 naira used, a return of 2 naira, 12 kobo (2.12) was gained. At Omupo, the IWDC3 SYN × 21 White DT STR SYN-DT C1 variety produced the highest yield of 2630 kg/ha and gave the highest profit of ₦79,650 and highest returns to investment of 205%. This implies that about two naira five kobo (2.05) was gained for every one naira used for production. In general, the DTMA varieties had higher profitability. The DTMA varieties had better yield than the farmers' variety at all locations of the women group trials.

5.4.4 Reasons for Preference

The yellow colour of seed was the most preferred characteristic at all the locations. The yellow maize was preferred whether it was the improved variety or farmer's variety. This is due to the fact that the farmers claimed that the yellow maize was marketable and is of high demand in the market as well as it commands a better price in the market than the white maize. It is also believed that the yellow maize is more nutritious than the white maize. Other reasons for preference include cobs with full grains, big seed, big cobs, DT, big leaves, greenish leaf colour, strongly matured cobs, produce multiple cobs, tall stalks or good stand, high vigour and attractive look.

5.5 Conclusion and Recommendations

The study revealed that the women farmer's varietal preference differs across locations. The women farmers ranked the DT maize varieties as the best at all locations. The profitability of the DT maize varieties also differs per location with the DT maize varieties having the highest profit and returns to investment. There was a long period of drought and delayed onset of rainfall (erratic rainfall) which affected the crops and resulted in lower yield than in previous years. However, the successes of thc on-farm trial encouraged the women farmers. It is therefore recommended that efforts should be made to involve women farmers in the varietal selection and testing procedure so as to ensure that the women farmers' preferences are incorporated in the development of agricultural technologies. This will help increase the farmers' yield and profitability from their production.

Finally, the study recommends that programmes and policies that will encourage women farmers' involvement in the development and testing of agricultural innovations should be implemented across the country in order to ensure food security and enhanced agricultural productivity.

Annexes

Table 5.1 Socio-economic characteristics of women farmers

Variables	Frequency	Percentage
Age		
<29 years	22	27.5
30–39 years	20	25.0
40–49 years	20	25.0
50–59 years	14	17.5
60 and above	4	5
Total	80	100
Mean age	43.11	
Marital status		
Single	4	5
Married	75	95
Total	80	100
Level of education		
No formal education	8	10
Primary education	42	52.5
Secondary education	20	25
Tertiary education	10	12.5
Total	80	100
Farm size		
1–5	80	100
6–10	0	0
Mean	1.92	

Table 5.2 Women farmer variety preference result for Lajiki

Variety	Low Preference (1)	Medium Preference (2)	High Preference (3)	Rank
TZEEI 95 × TZZEE 58	1	2	7	1st
TZEEI 81 × TZZEE 95	2	3	5	2nd
Farmer variety	9	1	0	3rd

Table 5.3 Women farmer variety preference result for Ballah

Variety	Low Preference (1)	Medium Preference (2)	High Preference (3)	Rank
TZEEI 95 × TZZEE 58	2	3	5	2nd
TZEEI 81 × TZZEE 95	0	2	8	1st
Farmer variety	4	5	1	3rd

Table 5.4 Women farmer variety preference result for Arandun

Variety	Low Preference (1)	Medium Preference (2)	High Preference (3)	Rank
EVDT 99	0	2	8	1st
99TZEE-Y-STR	2	3	5	2nd
Farmer variety	4	5	1	3rd

Table 5.5 Women farmer variety preference result for Omupo

Variety	Low Preference (1)	Medium Preference (2)	High Preference (3)	Rank
IWDC3 SYN × 21 White DT STR SYN-DT C1	0	1	9	1st
TZ COMP/ZDPSYN	0	3	7	2nd
Farmer variety	4	3	3	3rd

Table 5.6 Women farmer variety preference result for Isanlu Isin

Variety	Low Preference (1)	Medium Preference (2)	High Preference (3)	Rank
IWDC3 SYN × 21 White DT STR SYN-DT C1	0	2	8	1st
TZ COMP/ZDPSYN	1	3	6	2nd
Farmer variety	7	1	2	3rd

Table 5.7 Women farmer variety preference result for Mokwa

Variety	Low Preference (1)	Medium Preference (2)	High Preference (3)	Rank
2013 TZE-W DT STR	0	2	18	1st
2011 TZE-W DT STR SYN	2	4	14	2nd
Farmer variety	12	4	4	3rd

Table 5.8 Women farmer variety preference result for Kishi

Variety	Low Preference (1)	Medium Preference (2)	High Preference (3)	Rank
IWD C3 SYN/DT SYN-1-W	0	1	9	1st
2013 DTE STR-W SYN	1	2	7	2nd
Farmer Variety	6	2	2	3rd

Table 5.9 Economic analysis of the on-farm demonstration of women farmer group

Village	Variety	Yield (kg/ha)	Fertilizer costs (Naira)	Other expenses (Naira)	Total cost (Naira)	Gross returns (Naira)	Profit (Naira)	Return to investment
Lajiki	TZEEI 95 × TZZEE 58	1560	27,000	10,000	36,800	70,200	33,400	0.908
	TZEEI 81 × TZZEE 95	982	27,000	10,000	36,800	44,190	7390	0.201
	Farmer	797	27,000	10,000	36,800	35,865	−935	–
Ballah	TZEEI 95 × TZZEE 58	936	27,000	9270	36,270	42,120	5850	0.161
	TZEEI 81 × TZZEE 95	1539	27,000	9270	36,270	69,255	32,985	0.909
	Farmer	730	27,000	9270	36,270	32,850	−3420	–
Arandun	EVDT 99	2460	27,000	11,000	38,000	110,700	72,500	1.908
	99TZEE-Y-STR	2649	27,000	11,000	38,000	119,205	81,205	2.137
	Farmer	1281	27,000	11,000	38,000	57,645	19,445	0.512
Isanlu Isin	IWDC3 SYN × 21White DT STR SYN-DT C1	2630	27,000	11,600	38,600	118,350	79,750	2.066
	TZ COMP/ZDPSYN	2549	27,000	11,600	38,600	114,705	76,105	1.972
	Farmer	1021	27,000	11,600	38,600	45,945	7945	0.205
Omupo	IWDC3 SYN × 21White DT STR SYN-DT C1	2630	27,000	11,600	38,600	118,350	79,750	2.066
	TZ COMP/ZDPSYN	2549	27,000	11,600	38,600	114,705	76,105	1.972
	Farmer	1021	27,000	11,600	38,600	45,945	7345	0.19
Mokwa	2013 TZE-W DT STR	3460	27,000	11,000	38,000	155,700	117,700	3.097
	2011 TZE-W DT STR SYN	3649	27,000	11,000	38,000	156,105	118,105	3.108
	Farmer variety	2281	27,000	11,000	38,000	102,645	64,645	1.701
Kishi	2013 DTE STR-W SYN	4804	27,000	11,000	38,000	216,180	178,180	4.689
	IWD C3 SYN/DT SYN-1-W	4100	27,000	11,000	38,000	184,500	146,500	3.855
	Farmer variety	2280	27,000	11,000	38,000	102,600	64,600	1.7

Family labour cost was not assigned cost but similar labour time was used on each variety

References

Adeyemo, R., Kirk, M., & Ogunleye, A. S. (2015). Women Access to Land: The Compatibility of Property Rights on the Farming Activities of Women in Rice Producing Areas of Osun State, Nigeria. *International Journal of Research in Agriculture and Forestry, 2*(10), 34–42.

AFDB. (2015, August). *Economic Empowerment of African Women Through Equitable Participation in Agricultural Value Chains.* African Development Bank (AFDB) Report, p. 148.

Ajadi, A. A., Oladele, O. I., Ikegami, K., & Tsuruta, T. (2015). Rural Women's Farmers Access to Productive Resources: The Moderating Effect of Culture among Nupe and Yoruba in Nigeria. *Agriculture & Food Security, 4*, 26.

Ayinde, O. E., Abduolaye, T., Olaoye, G., & Akangbe, J. A. (2013). Gender and Innovation in Agriculture: A Case Study of Farmers' Varietal Preference of Drought Tolerant Maize in Southern Guinea Savannah Region of Nigeria. *Albanian Journal of Agricultural Science, 12*(4), 617–625.

Ayinde, O. E., Muchie, M., Adewumi, M. O., & Abaniyan, E. O. (2012). Risk Analysis of Gender in Innovation System: A Case Study of Production of Downy Mildew Maize Resistant Variety in Kwara State, Nigeria. *Obeche Journal, 30*(1), 459–465.

Ayinde, O. E., Abdoulaye, T., Takim, F. O., Oloyede, A. O., & Bankole, F. A. (2016). Economic Analysis of Onfarm Trial of Drought Tolerant Maize in Kwara State Nigeria: A Gender Approach. *Trakia Journal of Sciences, 14*(3), 287–293.

Bamire, A. S. Abdoulaye, T., Sanogo, D., & Langyintuo, A. (2010). *Characterization of Maize Producing Households in the Dry Savanna of Nigeria.* DTMA Country Report—Nigeria.

Beuchelt, T. D., & Badstue, L. (2013). Gender, Nutrition and Climate-Smart Food Production: Opportunities and Trade-Offs. *Food Security, 5*, 709–721.

CGIAR Big Facts: Evidence of Success Drought-Tolerant Maize Boosts Food Security in 13 African Countries. (n.d.). Retrieved February 2, 2017, from http://CCAFS.CGIAR.ORG/BIGFACTS/. CGIAR Research Program on Climate Change, Agriculture and Food Security (CCAFS).

CIMMYT. (2013, September). *The Drought Tolerant Maize for Africa Project.* DTMA Brief. http://dtma.cimmyt.org/index.php/about/background.

CIMMYT, & IITA. (2015). *CRP Maize Gender Strategy.* CGIAR Research Programme (CRP) on Maize Document, p. 34, CIMMYT, Mexico.

Enete, A. A., and Amusa, T. A. (2010) 'Challenges of Agricultural Adaptation to Climate Change in Nigeria: a Synthesis from the Literature', Field Actions Science Reports [Online], Vol. 4, 1–11.

FAOSTAT. (2010). *FAO Statistics On-line Database*. Retrieved from http://faostat.fao.org/.

Fisher, M., Abate, T., Lunduka, R. W., Asnake, W., Alemayehu, Y., & Madulu, R. B. (2015). Drought Tolerant Maize for Farmer Adaptation to Drought in Sub-Saharan Africa: Determinants of Adoption in Eastern and Southern Africa. *Climatic Change, 133*(July), 283–299.

Global Food Security Index. (2015). *The Role of Innovation in Meeting Food Security Challenges*. Special Report of the Economist Intelligence Unit Limited, p. 18.

Koyenikan, M. J., & Ikharea, V. E. (2014). Gender Access to and Control of Agricultural Resources in South Zone of Edo State, Nigeria. *Journal of Agricultural Economics, Extension and Rural Development, 1*(7), 138–145.

Lahai, B. N. (1994). An Evaluation of the Level of Female Participation in Cassava and Sweet Potato On-Farm Trials and Demonstrations in Sierra Leone. *Acta Hortic, 380*, 55–61.

Norman, D. W., Worman, F. D., Siebert, J. D., & Modiakgotla, E. (1995). *The Farming Systems Approach to Development and Appropriate Technology Generation*. Rome: Food and Agriculture Organization of the United Nations.

Olaniyan, A. B. (2015). Maize: Panacea for Hunger in Nigeria. *African Journal of Plant Science, 9*(3), 155–174.

Ortiz, R., Taba, S., Chávez Tovar, V. H., Mezzalama, M., Xu, Y., Yan, J., & Crouch, J. H. (2010). Conserving and Enhancing Maize Genetic Resources as Global Public Goods—A Perspective from CIMMYT. *Crop Science, 50*(January), 13–28.

Saito, K. A., Mekonnen, H., & Spurling, D. (1994). *Raising the Productivity of Women Farmers in Sub-Saharan Africa*. World Bank Discussion Paper 230, Washington, DC: The World Bank, p. 110.

Simonyan, J. B., Umoren, B. D., & Okoye, B. C. (2011). Gender Differentials in Technical Efficiency Among Maize Farmers in Essien Udim Local Government Area, Nigeria. *International Journal of Economics and Management Sciences, 1*(2), 17–23.

Soyemi, O. D. (2014). Women Farmers' Agricultural Information Need and Search Behaviour in North Central Nigeria. *Information and Knowledge Management, 4*(8). www.iiste.org.

6

Impact of Bioenergy Crop Adoption on Total Crop Incomes of Farmers in Northern Ghana: The Case of Jatropha Curcas

Lauretta S. Kemeze, Akwasi Mensah-Bonsu, Irene S. Egyir, D. P. K. Amegashie, and Jean Hugues Nlom

6.1 Introduction

Energy services have the potential to boost social and economic welfare of people. Access to energy is a crucial component of poverty alleviation, improving human welfare, and raising living standards (UNDESA 2005).

Most countries in sub-Saharan Africa rely on traditional biomass (crude oil, natural gas, and coal) as the primary energy source used and imported fossil fuels (IEA 2014). In the region, nearly 730 million people live in rural areas where they rely on traditional biomass for cooking (IEA 2014). In sub-Saharan Africa, energy demand grew by around 45%

L. S. Kemeze (✉) • A. Mensah-Bonsu • I. S. Egyir • D. P. K. Amegashie
University of Ghana, Accra, Ghana

J. H. Nlom
University of Maroua, Maroua, Cameroon

A. Shimeles et al. (eds.), *Building a Resilient and Sustainable Agriculture in Sub-Saharan Africa*, https://doi.org/10.1007/978-3-319-76222-7_6

between 2000 and 2012. In Africa, over 80% of electricity generated is from fossil fuels; about 620 million people do not have access to electricity (IEA 2014). Energy demand is predicted to double from 500 million tonnes oil equivalent (Mtoe) in the year 2000 to 1000 Mtoe in 2030 (Denruyter et al. 2010).

The heavy reliance on fossil fuels raises serious environmental issues such as depletion of non-renewable resources, ozone depletion, and global warming. According to Intergovernmental Panel of Climate Change (IPCC) (2007), global Greenhouse Gas (GHG) emissions should be reduced by 50–80% by 2050 to slow down global warming. This means that the use of fossil fuels for energy generation should be restrained. Even though Africa's GHG emissions account for less than 4% of the global GHG emissions, the continent is the most vulnerable to climate change effects such as droughts and flooding (World Bank 2009). This is because Africa is exposed to climate risks such as extreme droughts, flooding, and storms. In addition, its low adaptive capacity worsens the situation because the continent is characterized by high rates of poverty, financial and technological constraints, and heavy reliance on rain-fed agriculture.

Biomass is the dominant source of energy supply in Ghana. The country depends entirely on imports in order to meet oil requirements. The production of oil started with a capacity of 85,000 barrels of oil per day in Jubilee field (Abdulai 2013). In 2007, biomass energy consumption (wood fuel and charcoal) was about 11.7 million tonnes (Ministry of Energy 2010). It is used mainly for cooking, employing traditional inefficient technologies. Less than 10% of people use modern cooking fuels (improved stoves, kerosene, or liquefied petroleum gas) in the country (Ahiataku-Togobo and Ofosu-Ahenkorah 2009). In 2007, petroleum products and electricity consumption accounted for 1.955 million tonnes and 6269 GWh respectively (Ministry of Energy 2010). Biomass (fuelwood and charcoal) consumption in Ghana accounted for 64%. Petroleum products and electricity accounted for 27% and 9% respectively (Duku et al. 2011). However, according to IPCC (2007), combustion of fossil fuels contributes to global warming.

The energy sector faces some challenges in Ghana (Energy Commission 2006). These are related to the increase in energy demand, the potential imbalance between national energy production and indigenous sources of supply, inadequate investments in the energy sector, and overreliance on fuel imports and wood fuels. In order to address the issues of imbalance, low investments, and overreliance, Jatropha has been promoted as a panacea and promising feedstock for biofuels. However, Jatropha industry started in Ghana without any biofuel policies (Campion et al. 2012). In Ghana, prior to its introduction as a bioenergy crop, Jatropha was traditionally grown as gardens and a hedge or fence plant around homes in order to protect houses and fields against animals and sun exposure (Acheampong and Betey 2013). Jatropha was considered in Ghana for its ability to generate energy just recently, in 2005 (Boamah 2014). Its cultivation was promoted on marginal lands so as to not compromise food security (Boamah 2014). According to Brittaine and Lutaladio (2010), Ghana was predicted to be among the largest Jatropha producers in Africa by 2015. Projects related to Jatropha development started from 2005; by 2006, there were 17 biofuel projects in Ghana (Schoneveld et al. 2010). Several foreign companies (Agroils, Kimminic Estates, Jatropha Africa, Viram Plantation Limited, etc.) acquired large-scale land to produce both edible and non-edible crops for ethanol and biodiesel generation for exports (Dogbevi 2009). Large-scale Jatropha (100 hectares and more) development was highly criticized by Ghanaian NGOs for issues such as land grabbing and food insecurity as many of them were actually on fertile lands. Many of these large-scale Jatropha projects have failed. Of late, mainly participatory and small-scale Jatropha developments are ongoing in Ghana.

Jatropha seeds are not directly marketable in the open market. Farmers mostly sell their seeds to an NGO called New Energy. Previously, some farmers were selling the seeds to foreign investors under contract farming. However, these foreign investors are no more buying Jatropha seeds.

Income constitutes a key determinant of food security for poor people in rural areas since adequate income can help them afford appropriate food for their nutritional diet (FAO 2010; Faaij 2008). Jatropha can provide new income sources for farmers through Jatropha-generated activities such as seed selling. This supplementary income from Jatropha can impact the food security status of farmers helping them to afford food. According to the Government of Ghana (2010), the investment in land is predominant in the northern part of Ghana due to their high prevalence of food insecurity, poverty, and illiteracy. In addition, Northern Ghana agriculture accounts for more than 90% of household incomes and employs more than 70% of the population in the region.

The objective of this chapter is to measure the impact of Jatropha Curcas adoption on total crop incomes of farmers in Northern Ghana. Section 6.2 presents the theoretical framework and estimation technique. Section 6.3 provides data and descriptive analysis. Section 6.4 presents the empirical results. Finally, Sect. 6.5 provides the conclusions.

6.2 Theoretical Framework and Estimation Technique

6.2.1 Theoretical Framework: The Random Utility Framework

Following Hoque et al. (2015), a household's decision to adopt a bioenergy crop can be analyzed within a random utility framework. Let U_{hA} be the utility obtained by a household h from adopting Jatropha and U_{hN} the utility of non-adoption. Let Z_h be a vector of farm and household characteristics affecting bioenergy crop-adoption decisions and ε_h be the error term. According to the state of adoption, the household h utility is approximated as:

$$\begin{cases} U_{hA} = f\left(Z_h\right) + \varepsilon_{hA} \\ U_{hN} = f\left(Z_h\right) + \varepsilon_{hN} \end{cases} \tag{6.1}$$

A household will choose to adopt Jatropha only if the utility derived from adopting is greater than the utility from not adopting: $U_{hA} \succ U_{hN}$. Since these utilities are not observable, they can be expressed in the following latent structure model for adoption of bioenergy crop:

$$V_h^* = \beta Z_h + \varepsilon_h \tag{6.2}$$

$$V_h = \begin{cases} 1, V_h^* \succ 0 \\ 0, V_h^* \prec 0 \end{cases}$$

Where V_h is a binary indicator taking the value of 1 in the case the household adopts bioenergy crop and 0 otherwise.

The outcome variable (total crop incomes per hectare of the household) is considered as a linear function of the binary variable for bioenergy crop adoption along with a vector of some other explanatory variables (X):

$$Y_h = \lambda X_h + \gamma V_h + \mu_h \tag{6.3}$$

Where Y_h is the outcome variable, V_h is a binary variable for adoption, λ and γ are vectors of parameters to be estimated and μ is the error term. However, from Eq. 6.3, since γ measures the impact of bioenergy crop adoption (treatment variable) on total crop incomes per hectare (outcome variable), then, households should be randomly assigned to the group of adopters or non-adopters. However, technologies are rarely randomly assigned. Instead, new technology adoption usually occurs through self-selection. In other words, it translates the fact that in Eq. 6.3, μ is correlated with V or Z. Equation 6.2 which does not take into account the self-selection might lead to a biased estimation. The propensity score matching (PSM) is employed in this study in order to deal with selection bias.

6.2.2 Estimation Technique: Propensity Score Matching

Rosenbaum and Rubin (1983) defined the Average Treatment Effect (ATE) as follows:

$$\text{ATE} = Y_i^A - Y_i^N \tag{6.4}$$

where Y_i^A is the total crop income per hectare of household i that adopted and Y_i^N is the total crop income per hectare of household i that did not adopt. It is difficult to estimate the impact from Eq. 6.4. The issue is that either Y_i^A or Y_i^N is normally observed but not both of them for each household. What is normally observed is expressed as follows:

$$Y_i = K_i Y_i^A + (1 - K_i) Y_i^N \qquad \mathrm{K} = 0, 1 \tag{6.5}$$

where $K = 1$ represents the situation when the household i adopts Jatropha and $K = 0$ is the situation when the household has not adopted Jatropha.

The ATE can be re-specified as follows:

$$\begin{aligned} \mathrm{ATE} = P \cdot \left[E\left(Y_i^A / K = 1\right) - E\left(Y_i^N / K = 1\right) \right] + \\ (1 - P) \cdot \left[E\left(Y_i^N / K = 0\right) - E\left(Y_i^N / K = 0\right) \right] \end{aligned} \tag{6.6}$$

where P is the probability for a household to adopt Jatropha ($K = 1$).

Equation 6.6 is based on the assumption that the unobserved counterfactual of adopters if they had not adopted, $E\left(Y_i^N / K = 1\right)$, can be approximated by the one of non-adopters $E\left(Y_i^N / K = 0\right)$. Without that assumption, the estimation of Eq. 6.4 representing the ATE cannot be done because $E\left(Y_i^N / K = 1\right)$ is not observed. However, that procedure might highly result in a biased estimation because of the issue of selection bias. Indeed, the treated group (adopters) might not be statistically similar to the control group (non-adopters). Fortunately, the PSM approach of Rosenbaum and Rubin (1983) first reduces the pre-treatment characteristics of each household into one variable. Secondly, PSM uses the propensity score to match households with similar characteristics. Rosenbaum and Rubin (1983) defined 'propensity score' as the conditional probability of receiving a treatment given pre-treatment characteristics:

$$p(X) \equiv \Pr\{K = 1 / X\} = E\{K / X\};\ p(X) = F\{h(X_i)\} \tag{6.7}$$

where $F\{.\}$ is a normal or logistic cumulative distribution and X a vector of pre-treatment characteristics. An estimation of the propensity of

Jatropha adoption is run taking into account the restriction of the region of common support. After computing the propensity scores, the Average Treatment effect on the Treated (ATT) is estimated as follows:

$$\begin{aligned}\text{ATT} &= E\left\{Y_i^A - Y_i^N / K = 1\right\} \\ &= \mathrm{E}\left[\mathrm{E}\left\{Y_i^A - Y_i^N / K = 1, p(X)\right\}\right] \\ &= \mathrm{E}\left[\mathrm{E}\left\{Y_i^A / K = 1, p(X)\right\} - E\left\{Y_i^N / K = 0, p(X)\right\} / K = 1\right]\end{aligned} \tag{6.8}$$

ATT is performed using a single matching algorithm named Nearest Neighbor Matching with replacement. When there are few comparison units, matching with replacement allows one comparison unit to be matched more than once with each nearest treatment unit. However, matching without replacement forces the matching between the treatment group and the comparison group that is quite different in propensity scores. This enhances the likelihood of bad matches (increase the bias of the estimator). The quality of the matching is undertaken using a balance test called the mean absolute standardized bias. For each variable, the mean standardized difference is computed before and after matching as follows:

$$B(X) = 100 \frac{\overline{X_T} - \overline{X_C}}{\sqrt{\frac{V_T(X) + V_C(X)}{2}}} \tag{6.9}$$

where $\overline{X_T}$ and $\overline{X_C}$ are the sample means for the treated and control groups, $V_T(X)$ and $V_C(X)$ are the associated sample variances (Lee 2006). The bias reduction can be generated as follows:

$$BR = 100\left(1 - \frac{B_{\text{after}}}{B_{\text{before}}}\right) \tag{6.10}$$

Rosenbaum and Rubin (1985) recommended that the mean standardized bias after matching greater than 20% is perceived as an indicator of failed matching. In addition, according to Sianesi (2004), the balance test can be done comparing the pseudo R^2 and p-values from the propensity scores estimated before and after matching. After matching, there should not be any systematic differences in the distribution of covariates between adopters and non-adopters. As a result, the pseudo R^2 should be low. The test should be rejected after matching and not before.

6.3 Data and Descriptive Statistics

Data used in this study were collected from 400 farmers in the West Mamprusi and Mion districts of Northern Ghana using a questionnaire from September to October 2015. These districts were selected because of their involvement in Jatropha production. These districts are among the poorest in Ghana; hence, issues of innovation, crop diversification, or technology adoption for wealth creation which Jatropha promises become pertinent. For the purpose of this study, adopters are classified as farmers who planted Jatropha and still have it in their plots, while non-adopters refer to farmers who did not cultivate Jatropha. Focus group discussions consisted of meeting with community leaders and some Jatropha farmers in each district to gather preliminary information on the number of Jatropha growers in the communities and the size of their Jatropha plot. The survey used a structured questionnaire to collect data from the households on socioeconomic characteristics of households and information on Jatropha. From the preliminary study, the estimated number of Jatropha farmers in these two districts was 344 (256 farmers in West Mamprusi District and 88 farmers in Mion District). This information was used to calculate the minimum sample size.

Yamane's formula of sample size is used (Yamane 1967):

$n = \dfrac{N}{1 + N(e)^2}$ Where n is the sample size, N is the population size, and e is the error term. Assuming an error of 5% and a confidence interval of 95%, the following is obtained: $n = \frac{344}{1 + 344(0.05)^2} = 184.94$. The number of Jatropha adopters was increased to 200.

A stratified random sampling technique consisting of dividing the population into groups called strata and proceeding with a simple randomization was used to select the list of Jatropha farmers to be surveyed. The stratification was done at the district level. The first stage involved purposive selection of Jatropha growing districts in Northern Ghana. These are West Mamprusi District and Mion District. A total of 120 Jatropha farmers were randomly selected in West Mamprusi District and 80 in Mion District in order to have a fair representation of farmers in both districts based on the preliminary study.

The study needed a counterfactual to evaluate the effect of Jatropha farming adoption on key outcomes such as income. An equal sample size of 200 non-Jatropha farmers was used. The procedure to survey the non-Jatropha farmers was as follows: In each community where the Jatropha farmers were surveyed the equal number of non-Jatropha farmers was also surveyed. To choose a non-Jatropha farmer, a sample list of some non-Jatropha farmers in the community was collected from community leaders. A random list with three back-ups was then formed for each community to survey the non-Jatropha farmers.

Table 6.1 shows the distribution of respondents per district and community.

Figure 6.1 shows the map of the study area.

Table 6.2 describes the variables used in the study.

Table 6.3 shows the descriptive statistics of continuous variables used in the econometric models for the entire sample and the two sub-samples of non-adopters and adopters indicating the variable means and standard deviations. Adopters are distinguishable in terms of household characteristics such as age, farming experience, number of visits by extension services officers, and number of man-days labor hired.

On average, adopters allocate half a hectare to Jatropha cultivation. The mean age of farmers interviewed was about 43 years old. There is a significant difference in the age of adopters and non-adopters. On average,

Table 6.1 Distribution of respondents per district and community

District	Community	Adopters	Non-adopters
West Mamprusi District	Zagsilari	20	20
	Nasia	19	19
	Boamasa	20	20
	Janga	20	20
	Wungu	20	20
	Loagri	21	21
Mion District	Jimle	47	47
	Kpachaa	30	30
	Tuya	03	03
Total		200	200

Source: Authors

adopters are 45 years old while non-adopters are 41 years old. There is no statistically significant difference in the number of adult members in the household of both adopters and non-adopters. On average, the number of adult members in the household is five. There is no statistically significant difference in the years of education for farmers both adopters and non-adopters. On average, farmers spent two years of education. There is a significant difference in the years of farming experience at the level of 5% between the two groups. On average, Jatropha adopters have been farming for about 26 years while non-adopters for about 23 years. There is no statistical difference in farm size between the two groups and the average farm size is 3.75 hectares. On average, farmers own 3.45 hectares of land. There is no statistically significant difference in distance from the nearest agricultural market for the full sample. On average, the distance from home to the nearest agricultural market is 8.3 kilometers. There is a statistically significant difference at the level of 1% in the number of times farmers had access to extension services. Adopters had more access to extension services than non-adopters. On average, adopters had access to extension services 0.49 times compared to 0.29 for non-adopters during the 2014 cropping season. Adopters hired more labor than non-adopters. The mean number of hired man-days for adopters is 125 compared to 79.16 man-days for non-adopters. The difference is significant at the level of 5%. There is no statistical difference in the degree of risk attitude of the farmers. On average, the degree of risk attitude is 5.73.

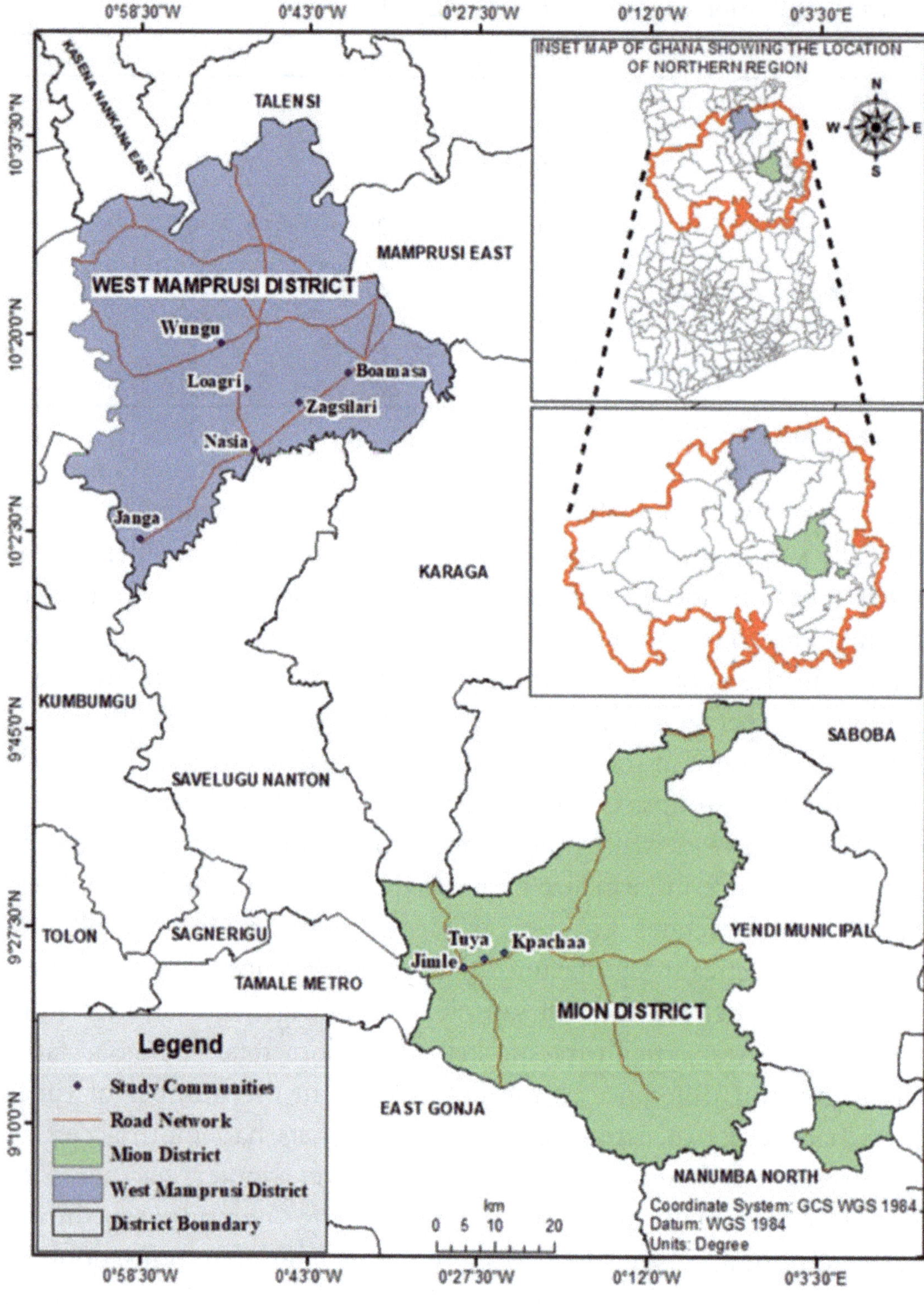

Fig. 6.1 Map of study area. Source: Authors

Table 6.2 Variables definition for Jatropha adoption

Variable	Type	Definition and measurement	Expected signs
Treatment variable			
Adopter	Dummy	1 = Grows Jatropha, 0 = Does not grow Jatropha	
Outcome variable			
Total crop incomes per hectare	Continuous	Income from all crops including Jatropha per hectare (in Ghana cedis)	
Independent variables			
Gender	Dummy	Gender of the head of household 0 = male, 1 = female	−
Age	Continuous	Age of the head of household (in years)	−
Education	Continuous	Level of education of the head of household in years	+
Number of adults	Continuous	Number of adult members of the household (count units)	+
Farming experience	Continuous	Farming experience of the head of household (years)	+
Farm size	Continuous	Farm size (hectares)	+
Extension services	Continuous	Number of times the household had access to extension services	+
Off-farm activities	Dummy	Engagement in off-farm activities 0 = No, 1 = yes	+
Livestock	Dummy	Livestock ownership 0 = No, 1 = yes	−
FBO	Dummy	Farmer based organization membership 0 = No, 1 = yes	+
District	Dummy	0 = Mion, 1 = West Mamprusi	−
Credit access	Dummy	Access to credit 0 = No, 1 = yes	+
Distance to market	Continuous	Distance from home to the nearest agricultural market (in km)	−
Size of land owned	Continuous	Size of land owned (in hectares)	+
Hired labor	Continuous	Number of man-days hired during 2014 cropping season	+
Irrigation	Dummy	Practice of irrigation 0 = No, 1 = yes	+
Risk attitude	Continuous	Degree of Risk attitude	+
Discount factor	Dummy	1 = preference for present, 0 otherwise	+

Source: Authors

Table 6.3 Descriptive statistics for continuous variables

	Adopters		Non-adopters		Total		
Variables	Mean	SD	Mean	SD	Mean	SD	t-value
Age	45.13	15.25	41.06	13.76	43.09	14.65	−2.80***
Number of adults	5.81	5.54	5.21	3.86	5.51	4.78	−1.25
Farming experience	26.46	15.53	23.32	14.20	24.90	14.94	−2.11**
Farm size	3.79	2.65	3.70	0.19	3.75	2.68	−0.32
Education	2.17	4.24	1.68	3.80	1.93	4.03	−1.20
Distance to market	8.11	7.08	8.47	7.32	8.30	7.19	0.51
Extension services	0.29	0.46	0.15	0.36	0.22	0.42	−3.39***
Size of land owned	3.56	2.46	3.35	2.61	3.45	2.53	−0.86
Hired labor	125.71	240.71	79.16	142.01	102.43	198.74	−2.35**
Risk attitude	5.93	2.69	5.53	2.66	5.73	2.68	−1.49

Source: Authors

Note: **, *** show significance at 5% and 1% levels, respectively

Table 6.4 shows the descriptive statistics of categorical variables used in the econometric models for the entire sample and the two sub-samples of non-adopters and adopters indicating the frequencies and percentages.

There is no statistically significant difference in gender of farmers between the two groups. On average, 85.5% farmers are male. Adopters are less engaged in off-farm activities (30%) compared to non-adopters (40.5%). There is a significant difference in farmer based organization (FBO) membership at 1%. The percentage of FBO membership is higher for adopters. A total of 45% of adopters are members of FBO compared to 22% for non-adopters. On average, 64% of respondents own live-stock. There is no statistically significant difference in access to credit between both groups. On average, only 19% of farmers had access to credit. The same for the discount factor, about 76% of farmers have a preference for the present. There is no statistically significant difference in irrigation practice; only 2% of respondents practiced irrigation.

Table 6.5 shows the descriptive statistics for total crop incomes per hectare of farmers for the whole sample and for male and female-headed households. There is a statistically significant difference at the level of 1% for the level of total crop incomes per hectare between adopters and non-adopters. On average, adopters have GHC 641.92 per hectare as total crop incomes while non-adopters have GHC 1243.41 per hectare.

Table 6.4 Descriptive statistics for categorical variables

		Adopters		Non-adopters		Total		
Variable	Category	No.	%	No.	%	No.	%	X^2 value
Gender	Male	166	83	176	88	342	85.5	2.02
	Female	34	17	24	12	58	14.5	
Off-farm act	Yes	60	30	81	40.5	141	35.25	4.83***
	No	140	70	119	59.5	259	64.75	
FBO	Yes	90	45	44	22	134	33.5	23.75***
	No	110	55	156	78	266	66.5	
Livestock	Yes	130	65	126	63	256	64	0.17
	No	70	35	74	37	144	36	
Credit access	Yes	38	19	38	19	74	19	0.00
	No	162	81	162	81	324	81	
Disc. factor	Yes	148	74	159	79.5	307	76.75	1.69
	No	52	26	41	20.5	93	23.25	
District	Mion	80	40	80	40	160	40	0.00
	WM[a]	120	60	120	60	240	60	
Irrigation	Yes	4	2	4	2	8	2	0.00
	No	196	98	196	98	192	98	

Source: Authors

Note: **, *** show significance at 5% and 1% levels, respectively

[a]West Mamprusi district

Table 6.5 Descriptive statistics for total crop incomes per hectare

Total crop incomes (GHC/ha)	Adopters		Non-Adopters		Total		
	Mean	SD	Mean	SD	Mean	SD	t-value
Whole sample	641.92	528.92	1243.41	1235.74	942.67	995.90	6.32***
Male	673.57	558.28	1206.98	1269.32	948.07	1024.18	4.98***
Female	487.39	314.68	1510.26	932.09	910.78	816.27	5.96***

Source: Authors

There is a statistically significant difference at the level of 1% for the level of total crop income for both male-headed and female-headed households. On average, male-headed household adopters have GHC 673.57 per hectare while male-headed household non-adopters have GHC 1206.98 per hectare. On average, female-headed household adopters have GHC 487.39 per hectare while female-headed household non-adopters have GHC 1510.26 per hectare.

Nonetheless, descriptive statistics cannot explain whether the observed difference in crop income per hectare between adopters and non-adopters for the whole sample and by gender is due to Jatropha cultivation. The noted differences in total crop income per hectare depending on the adoption status of the household might not be the result of Jatropha adoption but rather might be due to other factors, farm and farmers' characteristics, for instance. The impact analysis of Jatropha adoption on total crop incomes per hectare is preceded by the determination of the propensity scores for the treatment variable (adoption status of Jatropha).

6.4 Empirical Results

The impact analysis of Jatropha adoption on total crop incomes per hectare is preceded by the determination of the propensity scores for the treatment variable (adoption status of Jatropha).

6.4.1 Estimation of the Propensity Scores

A probit model is used to predict the probability of adopting Jatropha. The results of the propensity scores are reported in Table 6.6.

Several variables are statistically significantly associated with adoption of Jatropha. The number of times of access to extension services, the number of man-days hired, the FBO's membership, and the risk attitude of the head of household are positively associated with adoption. Access to extension services could play an important role in Jatropha adoption in Northern Ghana, for instance, in educating farmers in land use decisions concerning Jatropha. Jatropha is known as labor intensive; the ability of farmers to hire labor could increase its adoption. The coefficient of the risk attitude variable shows that the greater the degree of risk loving, the higher the probability of adopting Jatropha. Membership of FBO significantly increases the likelihood of Jatropha adoption. Indeed, FBO might assist farmers to manage the crop, find a market, and get access to loans.

The variables district and off-farm activities membership are negatively associated with Jatropha adoption. Being located in West Mamprusi District significantly reduces the likelihood of adopting Jatropha compared to Mion District. This is likely due to greater access to a potential

Table 6.6 Probit estimates of the propensity to adopt Jatropha

	Probit	
Variables	Coefficient	Standard error
Gender	0.286	0.205
District	−0.345*	0.181
Education	0.028	0.018
Age	0.014	0.007
Number of adults	0.026	0.016
Farming experience	−0.001	0.007
Farm size	−0.112	0.076
Extension services	0.200***	0.072
Off-farm activities	−0.299**	0.146
Livestock	0.097	0.146
Credit access	0.042	0.187
Distance to market	−0.001	0.011
Hired labor	0.002***	0.000
Size of land owned	0.063	0.081
FBO	0.583***	0.155
Risk attitude	0.086***	0.029
Discount factor	−0.178	0.179
Irrigation	−0.043	0.461
Constant	−1.176	0.335
Pseudo R^2	0.1269	
Log-likelihood	−242.08	
Observations	400	

Source: Authors

Note: *, **, *** show significance at 10%, 5%, and 1% levels, respectively

market in Mion District. The NGO New Energy is buying the seeds from the farmers in Mion district. Engagement in off-farm activities has a negative influence on the probability of Jatropha adoption. The current result might be due to the fact that farmers engaged in off-farm activities have less time and resources to engage in Jatropha cultivation.

Table 6.7 provides the distribution of the propensity scores.

For adopters, the estimated propensity scores vary between 0.06356 and 0.98953 with a mean of 0.50168. For non-adopters, it varies between 0.06356 and 0.89317 with a mean of 0.42073. The results suggest that the region of common support is satisfied in the interval [0.06356, 0.89317]. The consequence of this restriction is that observations falling outside this range of the region will be discarded from the analysis. As a result, nine observations have been removed from the

Table 6.7 Estimated propensity scores

Sample	Observations	Mean	SD	Minimum	Maximum
Whole sample	400	0.50168	0.20187	0.06356	0.98953
Adopters	200	0.58263	0.19964	0.11644	0.98953
Non-adopters	200	0.42073	0.16935	0.06356	0.89317

Source: Authors

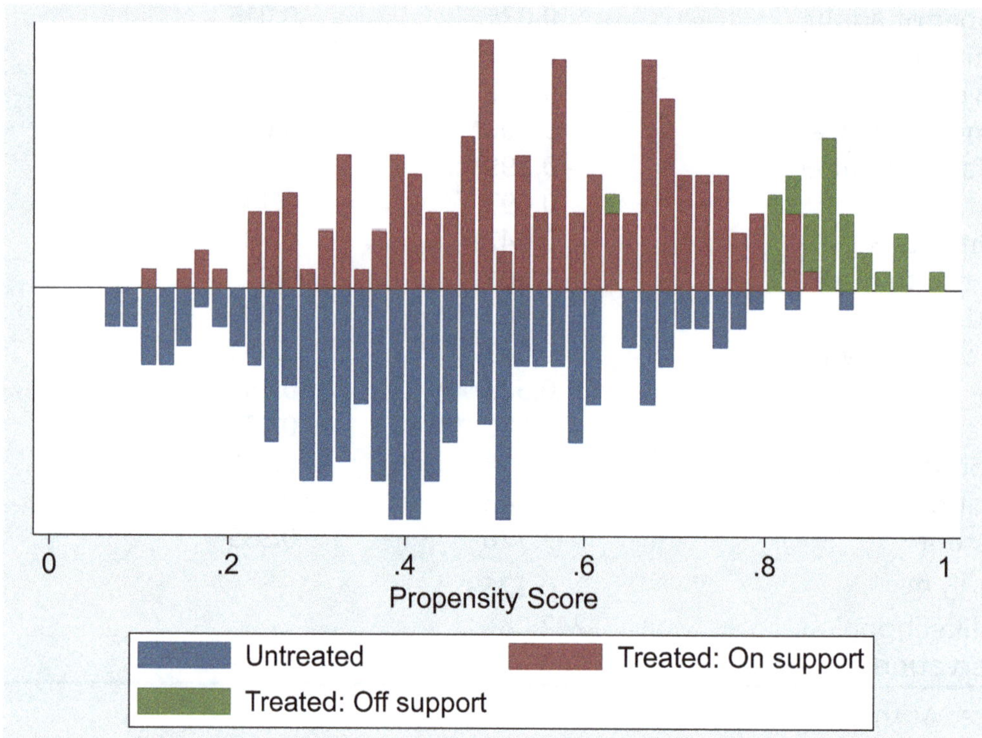

Fig. 6.2 Propensity score distribution and common support for propensity score estimation. Source: Authors

analysis. The common support condition is imposed in the regression models by matching in the region of common support only.

The distribution of the propensity scores and the region of common support before and after matching are represented in Fig. 6.2.

6.4.2 Estimation of Average Adoption Effect: Matching Algorithms

Table 6.8 reports the estimates of the average adoption effects estimated by Nearest Neighbor Matching with replacement for the whole sample

Table 6.8 ATT of Jatropha adoption on crop income

Matching algorithm: Nearest neighbor Matching with replacement	Sample	ATT	Number of treated	Number of control
Whole sample	400	−385.23*** (−3.05)	170	200
Male	342	−293.22** (−2.48)	143	176
Female	58	−624.23*** (−3.76)	15	24

Source: Authors
Note: **, *** show significance at 5% and 1% levels, respectively

and for male and female. The analysis is based on the restriction of the region of common support.

Table 6.8 shows that adoption of Jatropha significantly reduces total crop income per hectare of farmers. For the whole sample, the ATT estimate per hectare is negative (GHC −385.230). This is probably due to the fact that most farmers grow Jatropha on their fertile lands. Jatropha production is likely to compete for land with food production resulting in smaller areas cultivated and less food grown at the local level. The absence of an open market for Jatropha especially in the West Mamprusi district makes income generation from Jatropha very difficult. This results in reducing total crop incomes of farmers. The adoption of Jatropha significantly reduces the total crop income per hectare of both male-headed and female-headed households. The ATT estimate is GHC −293.22 for male and GHC −624.23 for female. Comparing the impact on male-headed and female-headed households, the results reveal that the participation in Jatropha cultivation affects more negatively female-headed than male-headed households. The reduction in total crop income per hectare is higher for female-headed than male-headed households. This can be explained by the fact that bioenergy crops such as Jatropha are inputs intensive (land, labor, water, fertilizers, and pesticides) and female farmers have traditionally limited access to inputs (Tauli-Corpuz and Tamang 2007). In Ghana, for instance, female farmers have very little access and control over resources due to patriarchy. They face challenges such as unequal access to land, finance and credit, and so on.

6.4.3 Indicators of Matching Quality Before Matching and After Matching

Table 6.9 provides the indicators of matching quality. It reveals the results of covariate balancing tests before and after matching. It can be seen that all indicators of matching quality before matching significantly exceed those after matching. After matching, the results show an insignificant likelihood ratio test supporting a rejection of the joint significance of covariates. In addition, after matching the results reveal a lower pseudo R^2. Indeed, the pseudo R^2 dropped from 0.127 to 0.051 after matching. After matching, there is also a reduction in absolute bias for overall covariates used to estimate the propensity score. Table 9 also reveals a mean standardized bias lower than 20% after matching as recommended by Rosenbaum and Rubin (1985). The standardized mean difference for overall covariates used in the propensity score around 15% before matching is reduced to about 13.3% after matching. This leads to a substantial reduction of the total bias of 11.33% through matching. All these statistics suggest that the specification of the propensity score is fairly successful in balancing the distribution of covariates between adopters and non-adopters. These results can then be used to assess the impact of Jatropha adoption among groups of farmers having the same observed characteristics.

Table 6.9 Matching quality indicators before and after matching for the whole population

Matching quality indicators	Before matching	After matching
Pseudo R^2	0.127	0.051
LR χ^2	70.35	23.89
$p > \chi^2$	0.000	0.159
Mean standardized bias%	15.0	13.3
Total % |bias| reduction	11.33	

Source: Authors

6.5 Conclusions

This study examined the adoption of Jatropha Curcas and its impact on total crop incomes of farmers in two districts of Northern Ghana. The PSM method was used to account for selectivity bias. The results suggested the presence of bias in the distribution of covariates between groups of adopters and non-adopters, indicating that accounting for selection bias is a significant issue. The results showed that Jatropha adoption reduces the total crop incomes per hectare of farmers. The study also highlighted the potential gender-differentiated impacts of Jatropha adoption on total crop incomes per hectare of farmers. The impact is worse for female-headed households compared to male-headed households, though the impact is negative for both. The ATE on the Treated estimates are GHC −385.230 per hectare for the whole sample, GHC −624.23 per hectare for female-headed households, and GHC −293.22 for male-headed household. Therefore, Jatropha cultivation might constitute a threat to farmers' crop incomes. There are a couple of reasons that can justify this finding, but the primary reason is the fact that Jatropha is cultivated in most cases on fertile lands and therefore conflicting with household staple and cash crops. The lack of market for Jatropha seeds is another reason which is worthy to highlight in this context.

The study recommends that the promotion of Jatropha cultivation should be properly regulated to avoid the massive conversion of fertile land used for crop production for Jatropha cultivation. There is a need to develop appropriate strategies and a regulatory framework to harness the potential economic opportunities from Jatropha cultivation while protecting rural people from converting part of their fertile lands to Jatropha cultivation at the expense of food crops. In this view, the Energy Commission (Ministry of Energy) should present the final bioenergy policy for Ghana in order for the country to move a step forward in the biofuel sector. The policy support is needed for improving income generation from Jatropha. Pro-women Jatropha development such as the promotion of Jatropha by-products (traditional soap and fertilizers) should be encouraged.

References

Abdulai, N. (2013). Ways to Achieve Sustainable Development in the Oil and Gas Industry in Ghana. *International Journal of ICT and Management, 1*(2), 107.

Acheampong, E., & Betey, B. (2013). Socio-Economic Impact of Biofuel Feedstock Production on Local Livelihoods in Ghana. *Ghana Journal of Geography, 5*, 1–16.

Ahiataku-Togobo, W., & Ofosu-Ahenkorah, A. (2009). Bioenergy Policy Implementation in Ghana. In Comptete International Conference 26–28 May 2009. Lusaka, Zambia. Retrieved from http://www.compete-bioafrica.net/events/events2/zambia/Session-2/2-2-COMPETE-Conference-Lusaka-Togobo-Ghana.pdf.

Boamah, F. (2014). Imageries of the Contested Concepts "Land Grabbing" and "Land Transactions": Implications for Biofuels Investments in Ghana. *Geoforum, 54*, 324–334. Retrieved from https://doi.org/10.1016/j.geoforum.2013.10.009.

Brittaine, R., & Lutaladio, N. (2010). *Jatropha: A Smallholder Bioenergy Crop the Potential for Pro-Poor Development*. Rome.

Campion, B., Essel, G., & Acheampong, E. (2012). Natural Resources Conflicts and the Biofuel Industry: Implications and Proposals for Ghana. *Ghana Journal of Geography, 4*(1), 42–64. Retrieved from http://www.ajol.info/index.php/gjg/article/view/110786.

Denruyter, J. P., et al. (2010). Bioenergy in Africa—Time for a Shift? *Sud Sciences & Technologies*, 145–158.

Dogbevi, E. K. (2009). Update: Any Lessons for Ghana in India's Jatropha Failure? *Ghana Business News*. Retrieved from https://www.ghanabusinessnews.com/2009/05/23/update-any-lessons-for-ghana-in-indias-jatropha-failure/.

Duku, M. H., Gu, S., & Hagan, E. B. (2011). A Comprehensive Review of Biomass Resources and Biofuels Potential in Ghana. *Renewable and Sustainable Energy Reviews, 15*(1), 404–415. https://doi.org/10.1016/j.rser.2010.09.033.

Energy Commission. (2006). *Strategic National Energy Plan (2006–2020)*. Ghana.

Faaij, A. (2008). *Bioenergy and Global Food Security*, Utrecht, Berlin: Wissenschaftlicher Beirat der Bundesregierung Globale Umweltverssnderungen (WBGU). Retrieved from http://www.wbgu.de/wbgu_jg2008_ex03.pdf.

FAO. (2010). *Bioenergy and Global Food Security: The BEFS Analytical Framework*. Rome.

Government of Ghana. (2010). *Medium-Term National Development Policy Framework: Ghana Shared Growth and Development Agenda (GSGDA), 2010–2013 Volume I Policy Framework Final Draft*. Accra. Retrieved September 9, 2017, from http://www.mofep.gov.gh/sites/default/files/docs/mdbs/2010/final_draft_mtdpf.pdf.

Hoque, M. M., et al. (2015). Producer Participation in Biomass Markets: Farm Factors, Market Factors, and Correlated Choices. *Journal of Agricultural and Applied Economics, 47*(3), 317–344.

IEA. (2014). *Africa Energy Outlook: A Focus on Energy Prospects in Sub Saharan Africa*. World Energy Outlook Special Report.

IPCC. (2007). Mitigation of Climate Change: Contribution of Working Group III to the Fourth Assessment Report of the Intergovernmental Panel on Climate Change.

Lee, W. (2006). Propensity Score Matching and Variations on the Balancing Test. In *3rd Conference on Policy Evaluation 27–28 October*. Mannheim (Germany).

Ministry of Energy. (2010). *National Energy Policy*, Republic of Ghana. Retrieved from http://old.sheltercentre.org/shelterlibrary/items/pdf/Ghana.pdf.

Rosenbaum, P. R., & Rubin, D. B. (1983). The Central Role of the Propensity Score in Observational Studies for Causal Effects. *Biometrika, 70*(1), 41–55.

Rosenbaum, P. R., & Rubin, D. B. (1985). Constructing a Control Group Using Multivariate Matched Sampling Methods that Incorporate the Propensity Score. *The American Statistician, 39*(1), 33–38.

Schoneveld, G. C., German, L. A., & Nutakor, E. (2010). Towards Sustainable Biofuel Development: Assessing the Local Impacts of Large-Scale Foreign Land Acquisitions in Ghana. In World Bank Land Governance Conference. Retrieved from http://siteresources.worldbank.org/EXTARD/Resources/336681-1236436879081/5893311-1271205116054/schoneveld.pdf.

Sianesi, B. (2004). An Evaluation of the Swedish System of Active Labor Market Programs in the 1990s. *Review of Economics and Statistics, 86*(1), 133–155.

Tauli-Corpuz, V., & Tamang, P. (2007). Oil Palm and Other Commercial Tree Plantations, Monocropping: Impacts on Indigenous Peoples' Land Tenure and Resource Management Systems and Livelihoods. In United Nations Permanent Forum on Indigenous Issues (UNPFII), 6th session, New York, 14–25 May 2007.

UNDESA. (2005). *The Millennium Development Goals Report 2005*. Retrieved from http://unstats.un.org/unsd/mi/pdf/mdg book.pdf.

World Bank. (2009). *Making Development Climate Resilient: A World Bank Strategy for Sub-Saharan Africa*. Washington, DC.

Yamane, T. (1967). *Statistics: An Introductory Analysis* (2nd ed.). New York: Harper & Row.

Part II

Addressing Climate Change Challenges

7

Profitability in a Sustainable Agricultural Production System: An Approach by the Soil and Water Conservation

Idrissa Ouiminga

7.1 Introduction

For several decades, Africa has suffered severe degradation of its natural resources, limiting the development of agro-sylvo-pastoral productions (Pontanier et al. 1995; Thiombiano 2000). Rising temperatures and changes in rainfall patterns have direct effects on crop yields and indirect effects due to changes in water availability for irrigation (IFPRI 2009). According to the Intergovernmental Panel on Climate Change (2014), yield reductions of 10–25% and even more could become commonplace by 2050.

The continent is experiencing difficult climatic conditions, relatively high population growth and a continuing decline in soil fertility. Repeated droughts and inadequate natural resource exploitation practices have resulted in the destruction of the vegetation cover and the exposure of the

I. Ouiminga (✉)
University Ouaga II, Ouagadougou, Burkina Faso

A. Shimeles et al. (eds.), *Building a Resilient and Sustainable Agriculture in Sub-Saharan Africa*, https://doi.org/10.1007/978-3-319-76222-7_7

soil to the weather (wind and rain). Since sub-Saharan agriculture is predominantly rain-fed, it is highly vulnerable to rainfall. (FAO 2010).

Thus, in the Sahelian regions, the areas of degraded and denuded soils are considerable, sometimes reaching significant proportions in northern Burkina Faso, with over 24% of the total agricultural area (Barro et al. 2005). For this country, 24% of arable land is severely degraded, and an average of 31% of annual rainfall is lost through runoff, posing a threat to food security in the medium and long term. Erosion of arable land, whether due to runoff or cultivation tools (Dibouloni 2004), is one of the factors contributing to declining agricultural yields (Roose et al. 1993).

To reverse this trend and achieve more sustainable patterns of exploitation, many actions to combat land degradation and desertification are needed. This includes assessing the effects of public or private investment in the management of natural resources through the transformation of production systems and the environment in the Sahel (Botoni and Reij 2009). This is why major financial and human investments have been made for the development and dissemination of soil and water conservation (SWC) techniques, which are considered as tools for soil protection and restoration. These techniques, by reducing erosion, contribute to the use of local resources (labor, stones, etc.) and waste reduction (animal and/or animal waste compost) and do not undermine the integrity of living beings.

The practice of soil and water conservation techniques is aimed at restoring soils and increasing yields and therefore affecting the environmental component. The social aspect is taken into account in the use of this practice because it constitutes a source of temporary income for the labor used. Moreover, at the social level, water and soil conservation techniques contribute to a reduction of migrations and a return of migrants (Ouedraogo et al. 2008).

However, nowadays it is important to include the notions of profitability in order to be able to take the appropriate decisions of management while taking into account the financial capacities of the targeted populations. However, even if the profitability of these techniques in terms of yield is proved, financially the question remains in the face of the lack of financial means available to the farmer, such as funds available for investment, lack of liquidity, work, and the earth. Indeed, costs related solely to labor are estimated between € 150 and € 230 per hectare (Barro et al. 2005). The evaluation of the financial profitability of soil and water conservation techniques in the municipality of Yalgo is the subject of our study.

Despite efforts by researchers over decades, the socio-economic impacts of investments in natural resource management (NRM) have been shown to be yield-positive (Ganaba 2005; Sawadogo H. 2008) but the returns on investments are varied, given the means invested for their implementation. Financial support through projects allows farmers to adopt them, while on an individual basis its facts are quite rare (Sanogo 2012). Baumgart-Getz et al. (2012) show that financial capacity is a basic factor of agricultural investment, which is why the producer expects a return on investment. Thus, the identification of the most profitable soil and water conservation technique in a Sahelian context is a tool for decision-making both for the promotion of sustainable agriculture in such climates and for the various anti-erosion projects and programs. Therefore, the determination of the social costs of production by SWC and the comparative evaluation of the benefits make it possible to identify the best technique in the Sahelian context.

Other authors have highlighted the use of practices related to the conservation of natural resources (soil, water, etc.). Gedikoglu and McCann (2007) who worked on conservation practices and Rodríguez-Entrena and Arriaza (2013) who worked on practices related to conservation agriculture all used the turnover realized by producers as indicators of wealth. The choice of investment indicator can also be the level of income (Mariano et al. 2012) or social capital (Gedikoglu et al. 2011); we opt for a more delicate measure, that of net profit.

SWC is the set of measures, which, while developing natural resources, tend to maintain (and if possible increase) the potential for production, the soil and water being the fundamental elements of these potentialities.

There are a large number of SWC techniques in our study area, the latter being mainly biological and physical.

Many organic techniques include organic fertilization and mulching.

Organic fertilization by compost or manure is fairly widespread. This is exclusively the application of raw manure or compost in cultivation plots (Dibouloni 2004). This spreading is also done by the parking of animals on the areas concerned. Sometimes parking is accompanied by contracts of fertilization between the breeders and the farmers.

Mulching consists of using mowed grass or crop residues that are spread on land to be recovered or improved during the dry season. This reduces the impact of water drops on the soil, reduces runoff, increases water infiltration into the soil, improves weed control, and the activity of microorganisms. The latter will favor the decomposition of straw or stems, thus contributing to the improvement of soil fertility (Ouédraogo 2005). However, the impact of mulching on yields, remains very low. Data collected at Donsin show an increase in yield of about 5% for mulching and 2% for burned mulch (Ouédraogo 2005).

Physical techniques include structures constructed or dug with the aim of creating obstacles to runoff and reducing soil erosion (Botoni and Reij 2009), including the basin, half-moons and stony cords.

The Zaï is an old peasant technique perfected by the various actors with the peasants. They are seed holes about 30 to 40 cm in diameter and 10 to 15 cm deep. The distance between the holes is 70 to 80 cm, which gives about 10,000 holes per ha. These holes are dug perpendicularly to the slope and staggered.

The half-moon is a practice of collection of runoff consisting of digging a basin in the form of a semicircle with a diameter between 2 m and 6 m and a depth of 15 cm to 20 cm (Kini 2007). A half-moon occupies a theoretical area of 1.57 to 14.13 m^2 and the number of half-moons per hectare is of the order of 312 to 417 according to the spacing between them (Ouédraogo et al. 2008).

The construction of stony cords or stone bunds is a semi-permeable structure consisting of two to three levels/rows of stones arranged in a contour (Lompo and Ouédraogo 2006). This technique slows down runoff so that it infiltrates more quickly.

In general, investments in natural resource management have important impacts and therefore contribute to increasing productivity and agricultural production. That result leads to increasing food security and improving of the population's income (Botoni and Reij 2009; Ouédraogo et al. 2008[1]). According to the same authors, on the social level, soil and water conservation techniques contribute to a reduction of migrations and a return of migrants. At the economic level, the quantification of impacts reveals that investments are very profitable, 37–107% for the zaï, 23% for the stony ropes and 145% for the half-moons.

There is therefore no doubt about the positive effects of SWC techniques, which are in the majority of cases supported by public investment projects or programs. If it is assumed that public investments or those through projects and programs are insufficient for the millions of farmers in developing countries, what about their profitability for financing from the farmer? What technique is there to ensure a good return on investment?

7.2 Methodology

There are several approaches to analyzing the financial viability of using natural resource management techniques. Economically, several methods of decision-making as to the justification of the opportunity cost of capital and the rate of social preference over time are possible.

The SWC techniques can be considered a club good, that means not a rival good but exclusive one (Samuelson 1954), insofar as the farmer who does not pay costs related to these techniques is excluded from their use. To do this, the farmer makes his reasoning rational: he carries out a benefit-cost analysis or a cost-benefit analysis through the determination of net present value (NPV) or internal rate of return (IRR).

The use of the NPV in such a case makes it possible to obtain an estimate of the net value of all the revenues generated by the use of these techniques over time.

$$\mathrm{VAN} = -I_0 + \Sigma \frac{R_t - C_t}{(1+i)^n} \pm \Sigma \frac{\mathrm{EXT}_t}{(1+i)^n} + \frac{\mathrm{Vd}}{(1+i)^n} = -I_0 + \Sigma \frac{\mathrm{CF}_t}{(1+i)^n}$$

where I_0 is the economic cost of the initial investment; R_t the exploitation Income; And economic operating costs; EXT_t positive or negative externalities; I the discount rate; Vd the residual value of the project; CF_t the cash flow of the investment; t the year of the project and n the project horizon.

As for the IRR (Ouédraogo et al. 2008), it gives the rate for which the NPV is zero. It is as follows:

$$\text{VAN} = 0 \Leftrightarrow I_0 = \sum \frac{\text{CF}_t}{(1+\text{TRI})^n}$$

One of the requirements of these two tools is the determination of the discount rate that is essential to their use so that an incorrect estimate of this rate leads to the obtaining of biased indicators. Moreover, the IRR would be relevant only if it is higher than the bank borrowing rate, which is not very evident in the context of developing countries. As financial risk is high, financial institutions take precautions and even discourage borrowing for such investments (Abramovay 2002).

Thus, the approach we use is the marginal rate of return (MRR) used by Crawford et al. (1991), Bourdon (1994) for similar studies.

The objective of this method is to determine the cost-effectiveness of different methods of restoring degraded lands. This analysis therefore aims to contribute to the formulation of recommendations that the farmer can adopt. Its application uses data collected over several years in real situations and meets the concept of sustainability. Indeed, it combines the economic aspects through the search for the treatment, giving the highest net benefit, the environmental aspects through treatments allowing a better management of the natural assets and social aspects through the workforce mobilized for the implementation and the sedentarization of the populations that it can bring.

The stages of economic analysis of SWC trials consist of four main parts:

- Preparation of the partial budget for each treatment
- The determination of the "higher" treatments whose profitability justifies the adoption by the farmer
- The calculation of the MRR for each "higher" treatment

- The determination, among the treatments considered to be sufficiently profitable, of which one seems to be the most interesting given the means available to the farmer and his objectives not yet taken into account in the analysis

7.2.1 The Preparation of the Partial Budget

In partial budgets, the net benefit of change is evaluated from current practices to recommended practices. For this analysis, we use costs and prices prevailing in the local market to estimate costs and revenues corresponding to the level of a given technical innovation.

7.2.2 The Determination of "Higher" Treatment

The identification of higher treatments is the first part of a marginal analysis. The goal of this task is to eliminate the lower treatments from the subsequent marginal analysis. Treatment is dominated or inferior when there is at least one other treatment with a higher net benefit for lower or equal loads. Treatment is therefore non-dominated, or superior, when there are no other options offering a higher net benefit for less than or equal loads.

7.2.3 The Calculation of the Marginal Rate of Return

The marginal rate of return for all treatments is calculated as the ratio (in percentage) of additional net income to the incremental costs associated with the adoption of an increasing level of input. The term "marginal" refers to the difference between the value of a given treatment and that of the lowest-ranking treatment; it is a ratio of variation to the margin. The marginal rates of return are compared with the target rate to identify satisfactory treatments. Treatment that meets the target rate is selected with the highest net benefit. We continue to change to another level of input provided that the MRR is above the target rate. In other words, the marginal rate of return indicates where expenditure ceases to provide a satisfactory increase in income, expressed as a percentage of invested funds.

7.2.4 Choice of Preferred Treatment

This step consists of choosing the treatment with the highest net profit and a MRR equal to or higher than the target rate.

7.2.5 Treatment

As presented in the problem, six combinations of treatments for soil and water conservation were applied compared to a control site without any treatment. The cultures produced are millet of variety IKMP5 and sorghum of variety Kapelga. The different treatments are as follows:

T1: producer's practice
T2: SR + Zaï + organic fertilizer (SR + Zaï + fo)
T3: SR + HM + organic fertilizer (SR + HM + fo)
T4: SR + Zaï + organic fertilizer + Urea (SR + Zaï + fo + Urea)
T5: SR + HM + organic fertilizer + Urea (SR + HM + fo + Urea)
T6: SR + Zaï + organic fertilizer + Urea + NPK (SR + Zaï + fo + Urea + NPK)
T7: SR + HM + organic fertilizer + Urea + NPK (SR + HM + fo + Urea + NPK)

SR = Stony ropes; HM = Half-moon; NPK = Nitrogen, Phosphorous, Potassium

7.3 Study Area and Data

7.3.1 Study Area

Our study focuses on testing SWC in Burkina Faso in Yalgo commune, more than 200 km north of the capital Ouagadougou. With an estimated density of 74.73 inhabitants/km^2 in 2014 (NISD 2015[2]), the municipality has a predominantly young population with a gender distribution of about 51% women and 49% men.

The basic activity is mainly agricultural with a predominance for food crops. In general, agriculture in the area is subject to severe land degradation, the grown areas are fragmented and their productivity is low. The crops commonly practiced by the populations are cereals such as millet, sorghum, corn, vouandzou.

Yalgo is one of the localities selected by the Institute of the Environment and Agricultural Research (INERA) as part of the project to improve water management in rain-fed systems to ensure food security in Burkina Faso (Improved water management in systems/AGES). The rainfall is typical of the Sudano-Sahelian climate, it is between 400 mm and 600 mm per year and the municipality has a single permanent watercourse. In this commune, the project covers four villages (Yalgo, Kario, Mamanguel and Taparko). The natural environment has a difficult context in the management of natural resources due to the arid and very hot climate.

7.3.2 Data

The data used for the work are mainly primary data. They range from 2014 to 2016 and cover the four villages of the municipality of Yalgo.

The characteristics of the primary data are summarized in Table 7.1.

- **Primary data**

The support we used to collect the primary data is the questionnaire. For the collection of primary data, a questionnaire allowed us to carry out a survey of 45 producers who took part in the AGES/INERA project. This survey takes into account the socio-personal, economic and institutional characteristics of the producers. It situates us on the different costs and revenues relative to the different technical options for water and soil conservation in order to determine their profitability.

It should be noted that all producers do not have access to credit, have received training in techniques for recovering degraded land, own their cultivated land, and almost all have access to the market.

Table 7.1 Description of primary data characteristics

	Variables	Modalities	Size	%
Socio-personal characteristics	Sex	0 = man	43	96
		1 = woman	2	4
	Perception	0 = bad	0	0
		1 = good	45	100
	Education	0 = no	43	96
		1 = yes	2	4
	Age	Average = 48	Min	25
			Max	71
	Active persons	Average = 5	Min	1
			Max	7
Economic characteristics	Market access	0 = no	1	2
		1 = yes	44	98
	Secondary activities	0 = no	0	0
		1 = yes	45	100
	Agricultural material	0 = no	38	84
		1 = yes	7	16
	Exploited area	Average = 4.9 ha	Min	2 ha
			Max	8.5 ha
Institutional characteristics	Credit access	0 = no	45	100
		1 = yes	0	0
	Training	0 = no	0	0
		1 = yes	45	100
	Land tenure	0 = no	0	0
		1 = yes	45	100
	Member of association	0 = no	45	100
		1 = yes	0	0

Source: Yalgo Surveys 2014, 2015

- **Secondary data**

For yields, we used the secondary data collected by the AGES project (2014–2016). Data on the costs of implementing SWC technologies (zaï, demi-lunes and stony rocks)[3] are mainly obtained from the Special Program for the Conservation of Water and Soils/Agroforestry (CES/AGF). The price of the various speculations (millet, sorghum) applied in the framework of project are from the cereals market information system of 2015.

7.4 Results and Discussions

7.4.1 Partial Budgets and Higher Treatments

As indicated in the methodology, the preparation of partial budgets is the first step in the process. They are set out in Table 7.2.

It is noted in Table 7.2 that treatment with the highest yield (T6) is characterized by the highest net benefit. It is also noted that peasant practice (T1) offers a higher yield than the T3 treatment. Also T1 (practice without arrangements) makes it possible to have a net profit greater than that of T3, T5 and T7. This may be explained, on the one hand, by the fact that the T3, T5 and T7 techniques have very high loads and insufficient yields. On the other hand, this phenomenon could be explained by the fact that the half-moon technique is less adapted to the culture of millet compared to the technique of the zai in the commune of Yalgo.

The partial budget of soil and water conservation techniques under Sorghum also shows that the technique with the highest yield (T7) also has the greatest net benefit. However, net income is not always proportional to performance. Indeed, it can be seen that peasant practice (T1), although performing below the T2 and T3 techniques, offers a higher net benefit than the latter. This may be because techniques T2 and T3 involve more loads than T1.

The identification of higher salaries, that is, salaries for which there is no other option offering a higher net profit for lower or equal charges, is made from the comparative results of profits and Costs.

Table 7.4 shows that the T3, T5 and T7 technologies are dominated because T1 allows a higher profit at a lower cost. This analysis of dominance under millet culture shows that only T2, T4 and T6 technologies can be considered as promising in terms of farmer practices (T1).

It is apparent from Table 7.5 that under sorghum cultivation only the T1, T5 and T7 technologies are superior.

By continuing the determination of the higher treatments between the two speculations, the results are presented to Table 7.6.

The comparative analysis between the higher treatments under sorghum and millet culture reveals that only the cultivation of millet has

Table 7.2 Partial budget for water and soil conservation techniques under cultivation of one hectare of Millet

	Treatments						
Topics	T1	T2	T3	T4	T5	T6	T7
Average yield (kg/ha)	620	2206	550	3197	650	3684	1100
Average yield readjusted (kg/ha)	558	1985	495	2877	585	3315	990
Production value (FCFA)	128,898	458,627	114,345	664,656	135,135	765,904	228,690
Monetary variable cost (FCFA)							
Cost per unit of organic fumure	0	6000	6000	6000	6000	6000	6000
Cost per unit of Urea	0	0	0	18,000	18,000	18,000	18,000
Cost per unit of NPK	0	0	0	0	0	25,000	25,000
Total variable cost	0	6000	6000	24,000	24,000	49,000	49,000
Non-monetary variable cost							
Cost of realization of stone	0	77,130	77,130	77,130	77,130	77,130	77,130
Cost of realization of zaï	0	29,600	0	29,600	0	29,600	0
Cost of realization of HM	0	0	27,200	0	27,200	0	27,200
Total opportunity cost	0	106,730	104,330	106,730	104,330	106,730	104,330
Total variable cost	0	112,730	110,330	130,730	128,330	155,730	153,330
Net benefit	128,898	345,897	4015	533,926	6805	610,174	75,360

Source: Survey 2015
NPK = 400 CFA/kg; Urea = 360 FCFA/kg; FO = 1.2 FCFA/kg, 231F/Kg of millet
FCFA - Franc of the Financial Communities of Africa

Table 7.3 Partial budget of SWC techniques under one hectare of Sorghum

	Treatments						
Topics	T1	T2	T3	T4	T5	T6	T7
Average yield (kg/ha)	583	1100	900	1407	2150	2522	2800
Average yield readjusted (kg/ha)	524	990	810	1266	1935	2269.8	2520
Production value (FCFA)	91,823	173,250	141,750	221,603	338,625	397,215	441,000
Monetary variable cost (FCFA)							
Cost per unit of organic fumure	0	6000	6000	6000	6000	6000	6000
Cost per unit of Urea	0	0	0	18,000	18,000	18,000	18,000
Cost per unit of NPK	0	0	0	0	0	25,000	25,000
Total variable cost	0	6000	6000	24,000	24,000	49,000	49,000
Non-monetary variable cost							
Cost of realization of stone	0	77,130	77,130	77,130	77,130	77,130	77,130
Cost of realization of zaï	0	29,600	0	29,600	0	29,600	0
Cost of realization of HM	0	0	27,200	0	27,200	0	27,200
Total opportunity cost	0	106,730	104,330	106,730	104,330	106,730	104,330
Total variable cost	0	112,730	110,330	130,730	128,330	155,730	153,330
Net benefit	91,823	60,520	31,420	90,873	210,295	241,485	287,670

Source: Survey 2015

Table 7.4 Identification of higher treatments under millet crop

	Variable cost (FCFA)	Net benefit (FCFA)	Superior?
T1	0	128,898	Oui
T2	112,730	345,897	Oui
T3	110,330	4015	Non
T4	130,730	533,926	Oui
T5	128,330	6805	Non
T6	155,730	610,173	Oui
T7	153,330	75,360	Non

Source: Survey 2015

Table 7.5 Identification of higher treatments under sorghum crop

	Variable cost (FCFA)	Net benefit (FCFA)	Superior?
T1	0	91,822	Oui
T2	112,730	60,520	Non
T3	110,330	31,420	Non
T4	130,730	90,872	Non
T5	128,330	210,295	Oui
T6	155,730	241,485	Non
T7	153,330	287,670	Oui

Source: Survey 2015

Table 7.6 Comparison of higher treatments under millet and sorghum crop

	Total variable cost	Net benefit	Superior?
T1 (millet)	0	128,898	Oui
T2 (millet)	112,730	345,897	Oui
T4 (millet)	130,730	533,926	Oui
T6 (millet)	155,730	610,173	Oui
T1 (sorghum)	0	91,822	Non
T5 (sorghum)	128,330	210,295	Non
T7 (sorghum)	153,330	287,670	Non

Source: Survey 2015

advantages. Indeed, the cultivation of millet offers higher profits than the sorghum crop for lower or equal variable costs.

7.4.2 Analysis of Profitability

The first step is to calculate MRR as shown in Table 7.7. It is the ratio of marginal net profit to marginal variable costs, expressed in relative terms.

These results at the margin show that for a farmer who passes from treatment T2 to treatment T4 the marginal gains (1044) are greater than when passing from T1 to T2 and/or from T4 to T6. The slope of the dominant options curve reflects the same result when linking only the top treatments (see Fig. 7.1).

Thus a major result is that it is not the treatment with the highest net benefit (T6) that gives the highest MRR but rather the T4 treatment.

7.4.3 Choice of Target Rate and Choice of Preferred Treatment

For African countries, the value-cost ratio standard accepted by Food and Agriculture Organisation (FAO) is 2; that means an MRR of 100%. This implies for the producer at least a doubling of the gains in relation to his investments. By observing the different MRR, this condition is fulfilled in the various cases (Table 7.7).

Based on this, the best combination of soil and water conservation techniques is T6 treatment.

Table 7.7 Marginal profitability rate

	Total variable cost	Marginal variable cost	Net benefit	Net marginal profit	MPR
T6	155,730	25,000	610,173	76,247	305
T4	130,730	18,000	533,926	188,029	1044
T2	112,730	112,730	345,897	216,999	193
T1	0		128,898		

Source: Survey 2015

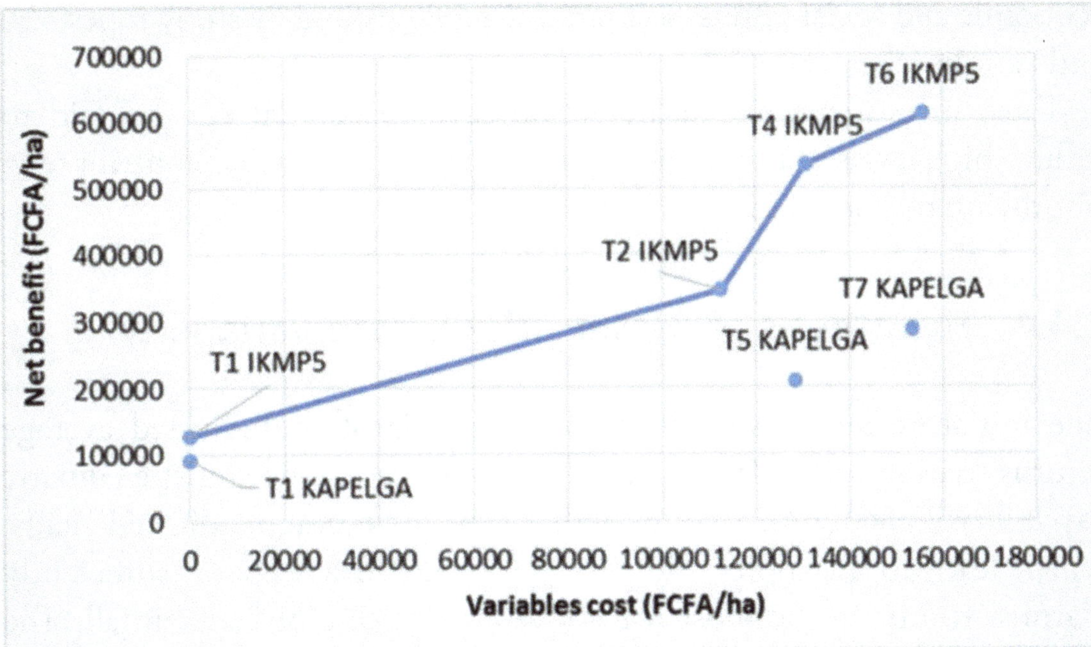

Fig. 7.1 Curve of dominant options. Source: Authors based on data of Survey 2015

All treatments with MRR equal to or above the target rate are satisfactory. Among the satisfactory treatments, the final choice of the treatment to be recommended will be determined by determining the treatment with the highest net benefit. Thus, with a view to profitability and in the context of better management of natural production assets, the choice will be made for T6 treatment (SR + Zaï + fo + urea + NPK) under millet culture. The material is a determining factor in the use of these techniques. The possession of small ruminants favored the adoption of zaï and the cattle were crucial for the adoption of stony ropes and "zaï and stony ropes". Not only is financial profitability guaranteed, but environmental recovery of the natural assets of land is an important achievement. It should be noted that the use of stony cords by the workforce that this requires entails a redistribution of income which is an important element of the social dimension.

Taking into account that producers have almost all access to the market and all own their growing areas, it can be assumed that the sale of millet, grown under T6, at the market price would be a benefit for producers. Such a result is termed sustainable because, in addition to the recorded economic and social results, soil restoration is ensured by stony ropes, Zaï and various fertilizers.

Thus, investments in natural resource management can be said to induce high levels of profitability, improve biodiversity and contribute to improving people's standard of living.

7.4.4 Hypothesis of Pessimistic Climate Variability

The initial estimates were made on the basis of the adjusted average returns so as not to inflate the results. In this case, a pessimistic climatic variability is taken into account with a 10% reduction in adjusted yields, which leads to an appreciation of the fallout in a scenario sufficiently alarmist to obtain the most realistic results in case of bad rainfall. The budgets estimated according to techniques and by speculation with pessimistic hypothesis are annexed to Table 7.8.

This table shows that the higher treatments in this scenario of poor rainfall are identical to those obtained in average rainfall conditions: T1, T2, T4 and T6 for millet and T1, T5 and T7 for sorghum.[4]

The comparison of the different higher treatments for both speculations is made to determine the best sustainable practice.[5]

The estimation in periods of unfavorable rainfall shows that the results achieved are less important from the financial point of view, but the observations remain the same as it is the combination T4 (SR + Zaï + organic fertilizer + Urea) which helps to obtain the highest marginal profit for the cultivation of millet.

7.5 Conclusion

A zone severely degraded by climatic factors (decrease in rainfall, winds, runoff) and at the edge of the Sahel, Yalgo is a Burkina Faso locality with a cereal deficit.

Various SWC techniques have been put into practice by research in order to contribute to increased yields and thus financial profitability.

Table 7.8 Budget for one hectare crop of millet and sorghum with pessimistic hypothesis

	T1	T2	T3	T4	T5	T6	T7
Millet							
Adjusted production	128,898	458,627	114,345	664,656	135,135	765,904	228,690
Pessimistic production	116,008.2	412,764.3	102,910.5	598,190.4	121,621.5	689,313.6	205,821
Total cost	0	112,730	110,330	130,730	128,330	155,730	153,330
Net benefit	**116,008.2**	**300,034.3**	**−7419.5**	**467,460.4**	**−6708.5**	**533,583.6**	**52,491**
Sorghum							
Adjusted production	91,823	173,250	141,750	221,603	338,625	397,215	441,000
Pessimistic production	82,640.7	155,925	127,575	199,442.7	304,762.5	357,493.5	396,900
Total cost	0	112,730	110,330	130,730	128,330	155,730	153,330
Net benefit	**82,640.7**	**43,195**	**17,245**	**68,712.7**	**176,432.5**	**201,763.5**	**243,570**

Source: Survey 2015

The main objective was to evaluate the most profitable opportunity for the application of water and soil conservation techniques.

Thus it is the combination of stony cords, water cuvettes, organic fertilizer, urea and NPK which makes it possible to obtain the highest benefit with millet cultivation. It is true that areas with low rainfall are recognized as suitable for growing millet; so, it is possible to improve agricultural performance not only financially but also physically in areas with a hard climate with better yields, protection of natural assets and an improved social fabric. It is also an alternative to adapting to climate change by sustainable and effective means.

Appendix

Table 7.9 Identification of higher treatments under millet and sorghum crops with pessimistic hypothesis

	Mil			Sorgho		
	Variable cost (FCFA)	Net benefit (FCFA)	Superior?	Variable cost (FCFA)	Net benefit (FCFA)	Superior?
T1	0	116,008.2	Oui	0	82,640.7	Oui
T2	112,730	300,034.3	Oui	112,730	43,195	Non
T3	110,330	−7419.5	Non	110,330	17,245	Non
T4	130,730	467,460.4	Oui	130,730	68,712.7	Non
T5	128,330	−6708.5	Non	128,330	176,432.5	Oui
T6	155,730	533,583.6	Oui	155,730	201,763.5	Non
T7	153,330	52,491	Non	153,330	243,570	Oui

Source: Survey 2015

Table 7.10 Comparison of higher treatments under millet and sorghum crops with pessimistic hypothesis

	Total variable cost	Net benefit	Superior?
T1 (millet)	0	116,008.2	Oui
T2 (millet)	112,730	300,034.3	Oui
T4 (millet)	130,730	467,460.4	Oui
T6 (millet)	155,730	533,583.6	Oui
T1 (sorghum)	0	82,640.7	Non
T5 (sorghum)	128,330	176,432.5	Non
T7 (sorghum)	153,330	243,570	Non

Source: Survey 2015

- **Calculation of the yield adjusted by treatment**

Yield reduction of 10% to account for differences in management, harvesting pattern and parcel size between trial and actual environment.

- **Determination of the price of the product**

Source: Cereal Market Information System (SIM). Price of millet = 231Fcfa/kg and Sorghum price = 175Fcfa/kg (in 2015).

- **Variables cost**

Monetary variable costs:

Table 7.11 Cost of input

Fertilizers	Quantity (Kg/ha)	Cost per hectare (F/ha)
Organic fertilization	5000	6000
Urea	50	18,000
NPK	62.5	25,000

Source: Survey 2015

Table 7.12 Cost of realization of zaï

Wording	Cost (FCFA/ha)
Material	4600
Workforce	25,000
Total cost per hectare	29,600

Source: Estimate of the CES/AGF Program, 2015

Table 7.13 Cost of realization of half-moons

Wording	Cost (FCFA/ha)
Material	7200
Workforce	20,000
Total cost per hectare	27,200

Source: Estimate of the CES/AGF du Programme, 2015

Table 7.14 Cost of realization of stony ropes

Wording	Cost (FCFA/ha)
Transport	29,545
Opening of trenches	6185
Tracking fees	5600
Small material	5800
Workforce	30,000
Total cost per hectare	77,130

Source: Estimate of the CES/AGF du Programme, 2015

Notes

1. Botoni and Reij 2009, Silent transformation of the environment and production systems in the Sahel: Impacts of public and private investments in the management of natural resources.
2. National Institute of Statistics and Development.
3. Data in Appendix (Tables 7.11, 7.12, 7.13, and 7.14).
4. See Table 7.9 in Appendix.
5. See Table 7.10 in Appendix.

References

Abramovay, R. (2002). Crédit rural et politiques publiques dans le sertão brésilien. *Revue Tiers Monde, 43*(172), 761–782.

Barro, A., Zougmoré, R., & Taonda, S. J. B. (2005). Mécanisation de la technique du zaï manuel en zone semi-aride. *Cahiers Agricultures, 14*(6), 549–559.

Baumgart-Getz, A., Prokopy, L., & Floress, K. (2012). Why Farmers Adopt Best Management Practice in the United States: A Meta-Analysis of the Adoption Literature. *Journal of Environmental Management, 96*, 17–25.

Botoni, E., & Reij, C. (2009). *La transformation silencieuse de l'environnement et des systèmes de production au Sahel: impact des investissements publics et privés dans la gestion des ressources renouvelables*. Centre for International Coopération (CIS) & Comité Permanent inter-Etats de Lutte contre la Sécheresse dans le Sahel (CILSS), Rapport, p. 63.

Bourdon, M. (1994). Faut-il craindre une agriculture capital-saving? *Économie rurale, 219*, 24–27. http://www.persee.fr/docAsPDF/ecoru_0013-0559_1994_num_219_1_4589.pdf.

Dibouloni, J. B. (2004). *Analyse de l'adoption des techniques du Zaï et des cordons pierreux dans les régions du Centre et du Centre-Sud (65p)*. Mémoire IDR, Université Polytechnique de Bobo Dioulasso.

FAO. (2010). Les implications du changement climatique pour le développement agricole et la conservation des ressources naturelles en Afrique. *Nature and Faune, 25*(1), 2026–5824.

Ganaba, S. (2005). Impact des aménagements de conservation des eaux et des sols sur la régénération des ressources ligneuses en zone sahélienne et nord soudanienne du Burkina Faso. *VertigO, 6*(2).

Gedikoglu, H., & McCann, L. (2007). *Impact of Off-Farm Income on Adoption of Conservation Practices*. Selected Paper at the American Agricultural Economics Association Annual Meeting, Portland, OR, p. 28.

Gedikoglu, H., McCann, L., & Artz, G. (2011). Off-Farm Employment Effects on Adoption of Nutrient Management Practices. *Agricultural and Resource Economics Review, 40*(2), 293–306.

IFPRI. (2009). *Changement climatique Impact sur l'agriculture et coûts de l'adaptation*. ISBN: 978-0-89629-536-0. Retrieved from http://www.ifpri.org/publication/climate-change-1.

IPCC. (2014). *Climate Change: Synthesis Report*. Contribution of Working Groups I, II and III to the Fifth Assessment Report of the Intergovernmental Panel on Climate Change [Core Writing Team, R. K. Pachauri, & L. A. Meyer (eds.)]. Geneva: IPCC. p. 151.

Kini, J. (2007). Analyse des déterminants de l'adoption des technologies de conservation des eaux et des sols dans le plateau central du Burkina Faso. Mémoire de DEA, Université de Ouagadougou, Burkina Faso.

Lompo, F., & Ouédraogo, S. (2006). *Rapport de l'étude pilote d'évaluation de l'impact des recherches en GRN en zone Sahélienne de l'Afrique de l'ouest 23*, p. 9.

Mariano, M. J., Villano, R., & Fleming, E. (2012). Factors Influencing Farmers' Adoption of Modern Rice Technologies and Good Management Practices in the Philippines. *Agricultural Systems, 110*, 41–53.

National Institute of Statistics and Development. (2015). *National Statistical Yearbook 2015*.

Ouédraogo, S. (2005). *Intensification de l'agriculture dans le plateau central au Burkina Faso: Une analyse possibilités à partir de nouvelles technologies*. Retrieved from http://dissertations.ub.rug.nl/files/faculties/eco/2005/s.ouedraogo/titlecon 2008.

Ouédraogo, S., Belemvire, A., Maiga, A., Sawadogo, H., & Savadogo, M. (2008). *Etude sahel Burkina Faso: Evaluation des impacts biophysiques et socio-*

économiques des investissements dans les actions de gestion des ressources naturelles au nord du plateau central du Burkina Faso. Rapport d'études, p. 94.

Pontanier, R., et al. (1995). *L'Homme peut-il refaire ce qu'il a défait?* (John Libbey Eurotext ed.pp. 179–188).

Rodríguez-Entrena, M., & Arriaza, M. (2013). Adoption of Conservation Agriculture in Olive Groves: Evidences from Southern Spain. *Land Use Policy, 34*, 294–300.

Roose, E., Kaboré, V., & Guenat, C. (1993). Le zaï: fonctionnement, limites et amélioration d'une pratique traditionnelle africaine de réhabilitation de la végétation et de la productivité des terres dégradées en region soudano-sahélienne (Burkina Faso). In *Spécial érosion: réhabilitation des sols et GCES. Cahiers ORSTOM*. Série Pédologie, 28 (2), 159–173. ISSN 0029-7259.

Samuelson, P. A. (1954). The Pure Theory of Public Expenditure. *The Review of Economics and Statistics, 36*(4), 387–389.

Sanogo, M. (2012). *Capitalisation des bonnes pratiques de gestion durable des terres pour l'adaptation à la variabilité et au changement climatique au Mali: analyse d'impacts agronomiques environnementaux et socio-économiques*. Centre régional AGRHYMET, mémoire, p. 84.

Sawadogo, H. (2008). Restauration des potentialités de sols dégradés à l'aide du zaï et du compost dans le Yatenga Burkina Faso. *Biotechnology Agronomy Society Environment, 12*(3), 279–290.

Thiombiano, L. (2000). Etude de l'importance des facteurs édaphiques et pédopaysagiques dans le développement de la désertification en zone sahélienne du Burkina Faso. *Thèse d'Etat, 1*, 209.

8

Land Tenure and Communities' Vulnerability to Climate Shocks: Insights from the Niger Basin of Benin

Boris Odilon Kounagbè Lokonon

8.1 Introduction

In sub-Saharan African (SSA) countries, the agricultural sector is expected to face serious difficulties due to climate change and variability (Fofana 2011). In these countries, agriculture is predominantly rain-fed, and consequently this sector is highly sensitive to climate change and variability. However, agriculture is the mainstay of the economy in most African countries, accounting for around 60% of Africa's employment and about one-quarter of the gross domestic product (GDP) (AfDB et al. 2015).[1] Farmers in these countries are mostly engaged in subsistence agriculture. Thus, the impacts of climate shocks and stresses are expected to translate into vulnerability, food and livelihood insecurity, and losses in human capital and in welfare (Davies et al. 2009).

It should be noted that climate-related shocks and stresses are not necessarily expected to lead to negative impacts on agriculture, because they are embedded in the practice of agriculture, and some farmers may develop

B. O. K. Lokonon (✉)
Université de Parakou, Parakou, Benin

A. Shimeles et al. (eds.), *Building a Resilient and Sustainable Agriculture in Sub-Saharan Africa*, https://doi.org/10.1007/978-3-319-76222-7_8

coping and risk management strategies (Davies et al. 2008). Moreover, the frequency of occurrence of climate shocks are expected to increase with climate change (IPCC 2013), and actions in terms of reducing the vulnerability and boosting the resilience of the population are needed. In addition, agriculture is recognized to play an important role in the structural transformation of Africa and in poverty reduction (AfDB et al. 2015).

Vulnerability has negative connotation. Thus, owing to that, resilience which originated in ecology (Holling 1973) is becoming influential in development economics. Resilience is not a pro-poor concept, and therefore it should be used with caution when trying to implement development actions (Béné et al. 2012). In addition, social protection is considered as an important factor in reducing poverty and vulnerability and in boosting resilience (Stern 2008; Davies et al. 2008, 2009; Solórzano 2016). Land tenure security is considered as part of social protection (Mahadevia 2011). Land tenure is relative to the conditions under which farmers hold and occupy the land (Schickele 1952). Therefore, agricultural productivity can be influenced by land tenure through the security (or investment) effect (Gavian and Fafchamps 1996; Yegbemey et al. 2013). For instance, Gavian and Fafchamps (1996) found that land tenure status is determinant in manure application between borrowed and owned fields in Niger; farmers have diverted manure toward the latter. Therefore, secure land tenure is increasingly considered as having an appropriate role in reducing the vulnerability of poor people to climate shocks (Jayne et al. 2003; Callo-Concha et al. 2013; Chagutah 2013). However, some factors such as lack of financial capital and access to technology can impede the potential of land tenure security in lessening vulnerability.

This chapter aims to assess the vulnerability of communities to climate shocks in the Niger basin of Benin and to analyze the extent to which land tenure influences vulnerability using the integrated approach and an econometric regression by taking advantage of two-period pseudo panel data. To date, there is limited understanding of the potential role of land tenure in reducing vulnerability of rural communities to climate shocks.

The remainder of the chapter is organized as follow. Section 8.2 presents the background and the conceptual framework. The specification of the vulnerability and resilience approach is presented in Sect. 8.3. Variables used and data sources are presented in Sect. 8.4. Section 8.5 presents the empirical results and discussion and Sect. 8.6 concludes.

8.2 Background and Conceptual Framework

Assessing the vulnerability of communities to climate shocks is important in identifying and characterizing actions toward strengthening resilience (Kelly and Adger 2000; Islam et al. 2014). Yet, existing literature suggests that individuals and communities that depend on highly climate-sensitive sector such as agriculture are vulnerable and less resilient to climate shocks. The existing literature is related to fishery systems (e.g., Islam et al. 2014), agricultural livelihoods (e.g., Brooks et al. 2005; Vincent 2007; Shewmake 2008; Deressa et al. 2008, 2009; Tesso et al. 2012; Etwire et al. 2013; Simane et al. 2016), and many sectors of the economy (e.g., Dixon et al. 2003; Dunford et al. 2015). However, to date none of them investigated quantitatively the extent to which land tenure affects vulnerability to climate shocks.

Vulnerability of communities to climate shocks is the propensity or predisposition they are to be adversely affected (adapted from IPCC 2014, p. 1775). The three components of vulnerability are exposure, sensitivity, and adaptive capacity of the communities. Exposure has an external dimension, while sensitivity, and adaptive capacity have an internal dimension (Füssel 2007). Exposure is the presence of communities in places and settings that could be adversely affected (adapted from IPCC 2014, p. 1765). Sensitivity refers to the degree to which communities are affected, either adversely or beneficially, by climate shocks (adapted from IPCC 2014, p. 1772). As for adaptive capacity, it is the ability of communities to adjust to climate shocks, to take advantage of opportunities, or to respond to consequences (adapted from IPCC 2014, p. 1758). Adaptive capacity encompasses five types of capital: physical, financial, human, natural, and social capital (Scoones 1998).

As mentioned above, resilience is becoming influential in development economics. Resilience is the capacity of communities to cope with climate shocks, responding or reorganizing in ways that maintain its essential function, identity, and structure, while also maintaining the capacity for adaptation, learning, and transformation (adapted from IPCC 2014, p. 1772). Vulnerability and resilience are related concepts (Turner 2010). Resilience influences adaptive capacity (Klein et al. 2003; Adger 2006). Both vulnerability and resilience recognize adaptive capacity, so they overlap through adaptive capacity (Berman et al. 2012). Some scholars view resilience as an integral part of adaptive capacity, while others consider adaptive capacity as a main component of vulnerability (Cutter et al. 2008). Moreover, there are scholars that see resilience and adaptive capacity as nested concepts within an overall vulnerability structure (Cutter et al. 2008). Cutter et al. (2008) viewed resilience and vulnerability as separate but often linked concepts. As for Turner (2010), vulnerability and resilience constitute different but complementary framings. Some researchers employ the term "resilience" to the coping capacity component of its framework, whereas others view vulnerability as an antonym of resilience (Turner 2010). Resilience can be considered as adaptive capacity in the case that it is used with an emphasis on society while also integrating environmental characteristics (Malone 2009). In this chapter, resilience is investigated through adaptive capacity, although resilience is not assumed as a synonym of adaptive capacity.

Scholars recognize the potential of social protection in reducing poverty and moving people into productive livelihoods (Davies et al. 2008). Social protection refers to all initiatives (public and private) which have the potential to reduce the economic and social vulnerability of poor, vulnerable, and marginalized groups, and social protection interventions can be classified as protective, preventive, promotive, and transformative measures (Devereux and Sabates-Wheeler 2004). Consequently, social protection can reduce poverty and move people into productive livelihoods, and is of paramount importance in helping the poorest to reduce their exposure to current and future climate shocks (Davies et al. 2008). Land tenure security is integral part of social protection (Mahadevia 2011).

In SSA, the livelihood of rural communities depends on land as key natural capital (Scoones 1998). Therefore, land tenure appears to be central to vulnerability and resilience research, although it is often overlooked (Berman et al. 2012; Chagutah 2013). Land tenure varies across households, communities, and individuals' characteristics such as gender and social groups. Higher levels of tenure security are considered to be associated with higher living conditions, human development achievements, economic status, and access to entitlements (Mahadevia 2011). Through land tenure security people have protection against their involuntary removal from their land without process of law (Mahadevia 2011). Secure land tenure militates for diversified livelihoods and favor investment in appropriate technologies and uptake of sound environment management practices (Economic Commission for Africa 2003; Chagutah 2013). However, land tenure security can lead to environmental degradation in rural areas where farmers operate under customary tenure (Chagutah 2013), and therefore exacerbate vulnerability to climate shocks.

Traditional common property systems constitute the basis of land tenure in rural Northern Benin (Callo-Concha et al. 2013). In Benin, ownership of property is acquired and transmitted by succession, donation, purchase, will, and exchange. Property can also be acquired by accession, incorporation, prescription, and other effects of obligations. In this chapter, the focus is on the institutional arrangements on land: the ways and arrangements through which farmers have access to land (Yegbemey et al. 2013).[2]

8.3 Specification of the Vulnerability and Resilience Approach

Following Lokonon (2017) conceptual framework and based on Sect. 8.2, vulnerability and resilience to climate shocks is assessed through an integrated approach using the indicator method. Vulnerability index is calculated as the net effect of adaptive capacity, sensitivity, and exposure.

$$Vulnerability = adaptive\,capacity - \left(exposure + sensitivity\right). \tag{8.1}$$

Weights are assigned to each indicator using the different weighting approach. Therefore, principal component analysis (PCA) (Pearson 1901) is used to attribute weight to the different indicators of the three dimensions of vulnerability to climate shocks. Factor scores from PCA are employed as weights to construct vulnerability indices for each village based on Eq. (8.1). Moreover, each indicator is normalized using the *z*-score standardization, and all the extracted factors from PCA are used due to the multidimensionality nature of vulnerability (Vincent and Cull 2014). Therefore, each factor is weighted by the explained variance.

The extent to which land tenure affects vulnerability is investigated through an econometric analysis. The vulnerability equation is specified as follows:

$$\upsilon_{it} = \beta_0 + Y_{it}\beta_1 + X_{it}\beta_2 + \vartheta_i + \gamma_{it} \tag{8.2}$$

where X_i is the set of variables belonging to the three dimensions of vulnerability apart from land tenure variables, Y_i is a variable reflecting land tenure security (the percentage of crop land which is owned by the farmers themselves), β_0, β_1, and β_2 are the vectors of the coefficients to be estimated, and $\vartheta_i + \gamma_{it}$ is the error term. All the variables used cannot be included in the regression for the sake of degree of freedom. Therefore, relevant regressors are chosen among the variables used to build vulnerability index through stepwise analyses. Panel specification tests are run to select the appropriate model (Baltagi 2008). Land tenure security is expected to negatively and significantly influence vulnerability to climate shocks. It should be noted that the variable capturing land tenure may be endogenous. Therefore, this chapter accounts for this likely endogeneity and use as instruments the departments in which the communities belong. Indeed, land tenure may vary with respect to the geographic settings.

Moreover, every model has to be tested for sensitivity and uncertainty. A Monte Carlo analysis (Metropolis and Ulam 1949) is performed to assess the uncertainty within the vulnerability index calculation model. Monte Carlo method calculates new results by relying on repetitive random sampling (Metropolis and Ulam 1949). The sensitivity of the vulnerability indicator to any variability in the input dataset is investigated through the change and omission of certain indicators.

8.4 Description of the Variables and Data

Variables that are used for the analysis capture the three aspects of the Intergovernmental Panel on Climate Change (IPCC) definition of vulnerability (exposure, sensitivity, and adaptive capacity). Adaptive capacity reflects the five capitals: physical, financial, human, natural, and social capital (Scoones 1998). According to Deressa et al. (2008), exactly how climate shocks affect income or any proxy of livelihood could be the best measure of sensitivity. However, it was not possible to find relevant data, so this research relies on the assumption that areas that experience climate shocks are subject to sensitivity due to loss in yield and thus in income.

Exposure variables capture changes in temperature and in rainfall from long-term mean (1952–2012), under the assumption that areas with higher changes in temperature and precipitation are most exposed to climate shocks. Table 8.1 presents the indicators used to assess vulnerability and resilience to climate shocks.

Two datasets are used for the analysis: 1998 small farmer survey data from the International Food Policy Research Institute and the Laboratoire d'Analyse Régionale et d'Expertise Sociale (IFPRI and LARES 1998) and the data of the survey which was implemented within the Niger basin of Benin in the 2012–2013 agricultural year. Regarding the later survey, three-stage sampling is used. First, communes were randomly chosen within each agro-ecological zone (AEZ), based on their number of agricultural households. Second, 28 villages were randomly selected within selected communes and last, random farm households within selected villages. AEZ V was disregarded, because only one of its communes is located within the Niger basin. The sample size is 545 agricultural households. The questionnaire used is composed of eight sections ranging from demographic information to household assets and basic services.

As for the 1998 small farmer survey, the households were selected using a two-stage stratified random sample procedure based on the 1997 Pre-Census of Agriculture. First, villages were randomly selected in each department, with the number of villages proportional to the volume of agricultural production. Second, in each village, nine households were randomly selected using the list prepared for the Pre-Census. The final sample size was 899 farm households in the country (153 farm households from 14 villages within the Niger basin of Benin). Regarding each data set, aggregation is done at village level using the weights attributed to each farm household. Moreover, additional information on the socio-economic infrastructures has been collected through an informal discussion with the village chiefs (number of primary, secondary and high schools, number of maternities, communal hospitals, district hospitals, dispensaries, clinics, and drinking water sources). In addition to the primary data, the research benefited socio-economic data from the Institut National de la Statistique et de l'Analyse Economique du Bénin (INSAE) and climatic data from the Agence pour la Sécurité de la Navigation Aérienne en Afrique et à Madagascar (ASECNA). The econometric analysis is on the 14 villages surveyed in 1998 and the 28 of 2012.

Table 8.1 Indicators used to assess communities' vulnerability and resilience to climate shocks

Vulnerability components	Indicators	Units	Hypothesized functional relationship
Exposure	Change in rainfall from the long-term mean	Percentage	Higher change = higher exposure
	Change in temperature from the long-term mean	Degree Celsius	
Sensitivity	Proportion of households that experienced flood over the last 20 years[a]	Proportion	Higher proportion = higher sensitivity
	Proportion of households that experienced droughts over the last 20 years[a]	Proportion	
	Proportion of households that experienced strong winds over the last 20 years[a]	Proportion	
	Proportion of households that experienced heat waves over the last 20 years[a]	Proportion	
	Proportion of households that experienced erratic rainfall over the last 20 years[a]	Proportion	
	Proportion of households that experienced heavy rainfall over the last 20 years[a]	Proportion	
	Proportion of households that experienced a change in planting date over the last 20 years[a]	Proportion	
	Proportion of households that experienced a decrease in yield over the last 20 years[a]	Proportion	

Adaptive capacity	Financial capital		
	Fertilizer-use value per household	Franc de la Communauté Financière Africaine (CFA F) F[b]	More financial capital = greater adaptive capacity
	Herbicide use value per household	CFA F	
	Insecticide use value per household	CFA F	
	Yearly income from agricultural off-farm activities per household	CFA F	
	Yearly income from non-agricultural off-farm activities per household	CFA F	
	Yearly income from cropping per household	CFA F	
	Yearly income from livestock per household	CFA F	
	Physical, institutional capital, and technology		
	Proportion of households within the communities that use plow	Percentage	Higher values = greater adaptive capacity
	Livestock value per household	CFA F	
	Amount of credit obtained per household	CFA F	
	Frequency of access to extension services per household	Frequency	
	Distance from dwelling to food market per household	Km	Higher distance = lower adaptive capacity
	Distance from dwelling to paved or tarred road per household	Km	
	Proportion of households within the communities that have access to electricity	Percentage	Higher values = greater adaptive capacity
	Household asset value per household	CFA F	Higher asset value = greater adaptive capacity
	Density of primary schools within the community	Quantitative	Higher values=greater adaptive capacity
	Density of secondary schools within the community	Quantitative	
	Density of high schools within the community	Quantitative	
	Density of maternities within the community	Quantitative	
	Density of municipality hospitals within the community	Quantitative	
	Density of district hospitals within the community	Quantitative	

(continued)

Table 8.1 (continued)

Vulnerability components	Indicators	Units	Hypothesized functional relationship
Human capital	Average household head formal education	Years	More human capital = greater adaptive capacity
Natural capital	Bush and valley bottom land-use size per household	Ha	More natural capital = greater adaptive capacity
	Irrigated land-use size per household	Ha	
Social capital	Proportion of households within the communities that belong to labor sharing groups	Proportion	Higher proportion = greater adaptive capacity
	Proportion of households within the communities that belong to farmers' organizations	Proportion	
	Amount of financial assistance per household	CFA F	Higher financial assistance and higher in-kind assistance = greater adaptive capacity
	Value of assistance in nature per household	CFA F	

Source: Author

[a]Proportion of households within the communities

[b]Currency of African Financial Community. In 2012, \$1 = CFA F 510.53

The Niger basin covers 37.74% of Benin, is located in the extreme north of the country between latitudes 11° and 12°30′ N and longitudes 2° and 3°20′40 E, and has an area of 43,313 km². Five AEZs out of the eight of the country are covered by the basin (wholly and partially). It covers 17 communes, both wholly and partially (12 communes wholly, and 5 partially). These communes belong to 3 departments: Alibori, Atacora, and Borgou. The agricultural sector in Benin employs 70% of the active population, and contributes 35% to the GDP, 75% to export revenue (République du Bénin 2014). The agricultural production is extensive, relies on family labor combined with limited use of improved inputs, production methods, and farm equipment. The country's agricultural trade is characterized by a weak performance, with a persistently negative agricultural trade balance.

8.5 Results and Discussion

8.5.1 Socio-economic Characteristics of the Communities and Environmental Attributes

The average percentage of farm households that used plows through animal traction within the communities amounted to 61% and 54% in 1998 and 2012, respectively. On average, the communities were poor in terms of income and asset ownership. The average yearly income per household within the communities from crop selling was CFA F 636,540.89 in 1998 and 1,423,760.69 in 2012. The value of the assets per household (except land) amounted to an average of CFA F 188,969.93 and 309,607.40 in 1998 and in 2012, respectively. Given that the subsistence and mixed crop-livestock production system was the dominant production system, livestock keeping was common among the surveyed communities. Livestock were used for consumption, traction, and manuring in farming, and as a means for cash income. The majority of cattle owned were for traction purposes. In terms of income from livestock, on average, a household within the communities earned CFA F 248,289.88 and 78,372.93 in 1998 and in 2012, respectively.

The farm households within the communities actively seemed to participate in off-farm activities to increase their livelihoods. The average yearly income per household from agricultural off-farm activities amounted to CFA F 11,063.49 and 30,596.35 in 1998 and in 2012, respectively. While the average yearly income from non-agricultural off-farm activities amounted to CFA F 96,856.92 and 293,713.73 in 1998 and in 2012, respectively.

Basic services and infrastructure were generally poor in the surveyed villages as is the case with the rest of the country. The communities had generally access to extension services through cotton production. However, in Malanville, a commune located at the vicinity of the Niger River, they had access to extension services through rice production. On average, the farm households had access 0.71 time to extension services in 1998 and 1.18 times in 2012. In fact, cotton production is organized in Benin through the farmers' organizations and 79% and 36.3% of the households were members of these organizations in 1998 and 2012, respectively. Access to health care was relatively low (low density of health infrastructures). The average amount of credit received per household within the communities amounted to CFA F 14,952.38 and 19,357.33 in 1998 and in 2012, respectively. Only 10% of the farm households within the communities had access to electricity in 1998 and 23% in 2012. Therefore, the percentage of households within the communities that have access to electricity had at least double between 1998 and 2012. However, they were too far away from paved or tarred roads, which meant that they did not have access to adequate roads even though the situation has been improved between the two periods.

Apart from the farmers' organizations, the farm households within the communities worked together through labor-sharing groups. Through labor-sharing groups, they alternated working on the farms of each member of the group. About one-third and a quarter of the farm households (31% and 24%) within the communities belonged to at least one labor-sharing group in 1998 and in 2012, respectively. The data reveal the existence of social capital in the basin. Indeed, the value of in-kind assistance per household amounted to CFA F 1902.95 in 2012, whereas the financial assistance per household within the communities amounted to 3178.57 and 3364.58 in 1998 and in 2012, respectively.

As for land ownership, in 1998, two types of tenure were found in the basin: owned land and others (use without paying any fee, and commune property). Land tenure security appeared to be high (at least 50% of crop land) within the communities except for one village (Donwari) which had 40.75% of owned land. The situation in 2012 differs relatively from that of 1998. Indeed, in 2012, there were lease and rented land in some communities (e.g., Bodjecali, Garou 1, Tintinmou Peulh, and Perma), although their level is low compared with owned land (at least 75% of crop land is under tenure security). During the last 20 years, the communities faced many climate shocks. Strong winds were the major climate shock that the communities faced over the last 20 years, followed by erratic rainfall, heavy rainfall, heat waves, floods, and finally droughts. The distribution of the shocks differs across villages.

8.5.2 Vulnerability and Resilience Levels of the Communities

Factor scores from the extracted components are employed to construct indices for adaptive capacity (financial capital, physical capital, institutional capital and technology, human capital, natural capital, and social capital), sensitivity, and exposure. The analyses help to understand the situation of 14 villages over time (in 1998 and in 2012) and 28 villages in 2012 (Tables 8.2 and 8.3). Higher values of the vulnerability indices depict less vulnerability, whereas lower values show more vulnerability. It is worth mentioning that on average both 1998 and 2012 were wet years. However, water excess was higher in 1998 than in 2012 (Figs. 8.1 and 8.2). The 1998 survey did not include the sensitivity to climate shocks and therefore, it was not possible to build the sensitivity index for this year. The situation of the villages has been improved except for Kossou, Kpbébéra, Gantiéco, Kota Monongou, and Moupémou.

Sirikou is the less vulnerable community in 2012, whereas the most vulnerable is Kota Monongou. Indeed, Sirikou is in the AEZ II and has a vulnerability index of 3.14. Kota Monongou is in the AEZ IV and has −2.48 as vulnerability index. In 2012, the range between sensitivity, exposure, and adaptive capacity of the two communities is 1.90, 0.59, and

Table 8.2 Vulnerability index and its components across villages

	Exposure		Sensitivity	Adaptive capacity		Vulnerability without sensitivity		Vulnerability
Villages	1998	2012	2012	1998	2012	1998	2012	2012
Bodjecali		−0.123	−0.937		−0.490		−0.367	0.570
Garou 1		−0.123	0.769		−1.123		−1.00	−1.769
Kassa	0.121	−0.123	0.451	−0.462	−0.597	−0.583	−0.474	−0.925
Toumboutou		−0.123	−0.863		0.538		0.661	1.524
Angaradebou		−0.230	0.275		1.791		2.021	1.746
Sonsoro Bariba		−0.230	−0.501		1.424		1.654	2.155
Donwari	−0.123	−0.230	0.362	0.719	1.379	0.842	1.609	1.247
Tankongou	−0.123	−0.230	−0.425	0.143	0.509	0.266	0.739	1.164
Kandifo Peulh	−0.123	−0.230	0.575	−0.119	1.107	0.004	1.337	0.762
Bouhanrou	−0.133	−0.069	−0.453	0.737	0.851	0.869	0.920	1.373
Tintinmou Bariba		−0.069	0.655		1.269		1.338	0.683
Tintinmou Peulh	−0.133	−0.069	0.004	−0.303	0.577	−0.170	0.646	0.642
Sirikou	−0.133	−0.069	0.585	2.101	3.066	2.234	3.135	2.550
Tepa (Gan Maro)		−0.008	−0.762		1.120		1.128	1.890
Kali		−0.008	−1.131		−0.624		−0.615	0.516
Serekale Centre		−0.008	−0.417		−0.358		−0.349	0.068
Kassakpere		−0.008	0.115		−0.816		−0.807	−0.922
Bembereke Ouest		−0.064	−0.605		−1.010		−0.946	−0.341
Kossou	−0.126	−0.064	0.416	0.934	−2.322	1.060	−2.259	−2.674
Kpebera	−0.126	−0.064	−0.156	0.573	−0.652	0.699	−0.588	−0.432
Kabanou	−0.126	−0.064	0.025	0.142	0.842	0.268	0.906	0.881
Makrou-Gourou		0.285	0.338		0.326		0.040	−0.297
Beket Peulh	0.185	0.285	0.718	−0.681	−0.292	−0.866	−0.577	−1.295
Gantieco	0.185	0.285	−0.081	−0.835	−1.257	−1.021	−.543	−1.462
Chabi Couma		0.285	0.220		−1.120		−1.405	−1.624
Kota Monongou	0.327	0.356	0.547	−1.972	−2.125	−2.299	−2.481	−3.028
Moupemou	0.327	0.356	−0.137	−0.976	−1.657	−1.303	−2.013	−1.877
Perma		0.356	0.415		−0.354		−0.710	−1.125

Table 8.3 Adaptive capacity components across villages

Villages	Financial capital		Physical, institutional capital, and technology		Human capital		Natural capital		Social capital	
	1998	2012	1998	2012	1998	2012	1998	2012	1998	2012
Bodjecali		−0.288		0.266		0.086		−0.299		−0.255
Garou 1		−0.086		−0.264		0.029		−0.771		−0.031
Kassa	−0.375	−0.489	0.073	−0.146	0.031	−0.052	−0.043	0.237	−0.148	−0.147
Toumboutou		0.360		0.230		0.007		−0.230		0.171
Angaradebou		0.669		0.045		−0.040		0.225		0.891
Sonsoro Bariba		0.562		0.455		0.013		0.301		0.093
Donwari	0.381	0.495	0.487	0.065	−0.219	0.164	0.124	0.191	−0.054	0.464
Tankongou	0.409	0.143	0.094	0.367	−0.005	0.102	−0.301	0.246	−0.123	−0.350
Kandifo Peulh	0.089	0.212	−0.138	−0.439	0.080	0.125	−0.097	0.229	−0.054	0.981
Bouhanrou	0.787	0.753	0.192	0.149	0.095	−0.083	−0.364	0.181	0.027	−0.149
Tintinmou Bariba		0.665		0.037		−0.032		0.174		0.425
Tintinmou Peulh	0.196	0.601	−0.079	0.154	−0.148	0.108	−0.161	−0.104	−0.111	−0.183
Sirikou	0.966	1.637	0.668	0.523	0.095	0.024	0.426	0.420	−0.054	0.463
Tepa (Gan Maro)		0.257		0.583		0.122		−0.026		0.184
Kali		−0.298		−0.103		−0.021		−0.017		−0.185
Serekale Centre		−0.294		0.109		−0.018		0.142		−0.297
Kassakpere		−0.201		−0.133		−0.116		−0.013		−0.353
Bembereke Ouest		−0.171		−0.232		−0.178		−0.038		−0.392
Kossou	−0.167	−1.130	0.367	−0.578	−0.091	−0.066	0.389	−0.089	0.436	−0.410
Kpebera	−0.065	0.038	−0.010	−0.364	0.095	−0.094	0.124	0.052	0.429	−0.344

(*continued*)

Table 8.3 (continued)

Villages	Financial capital		Physical, institutional capital, and technology		Human capital		Natural capital		Social capital	
	1998	2012	1998	2012	1998	2012	1998	2012	1998	2012
Kabanou	0.055	−0.210	−0.082	0.165	0.031	0.187	0.221	0.056	−0.083	0.645
Makrou-Gourou		0.308		−0.146		0.102		0.008		0.053
Beket Peulh	−0.433	−0.339	−0.158	−0.347	0.095	0.119	−0.170	−0.088	−0.013	0.363
Gantieco	−0.329	−0.270	−0.409	−0.382	0.023	−0.038	−0.115	−0.234	−0.005	−0.334
Chabi Couma		−0.532		0.098		−0.212		−0.044		−0.430
Kota Monongou	−1.002	−1.120	−0.553	−0.600	0.052	−0.010	−0.030	−0.090	−0.439	−0.305
Moupemou	−0.512	−0.652	−0.451	−0.443	−0.133	−0.060	−0.002	−0.207	0.122	−0.296
Perma		−0.630		0.930		−0.167		−0.213		−0.274

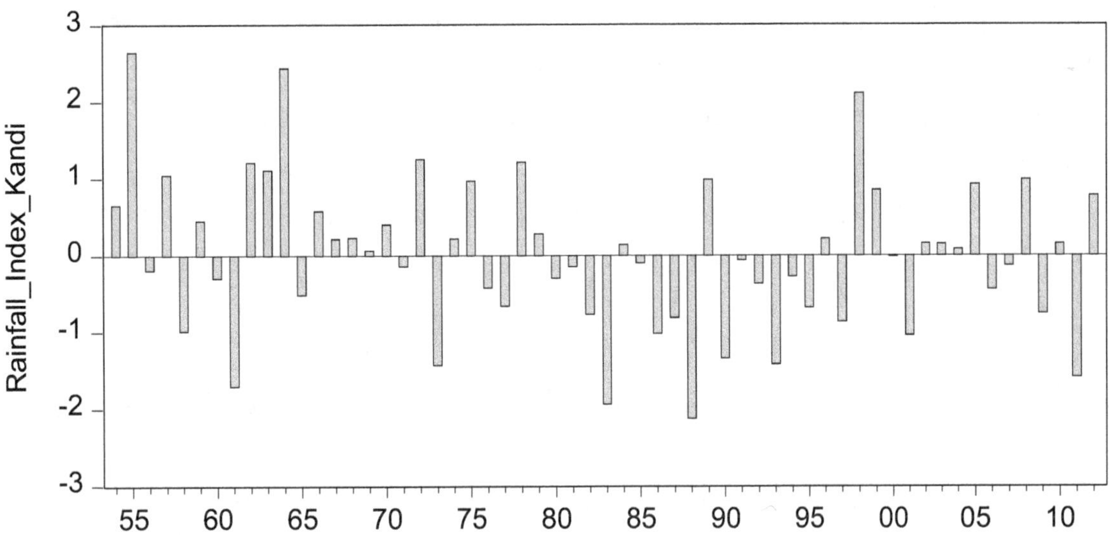

Fig. 8.1 Rainfall index evolution between 1954 and 2012 in Kandi. Note: Rainfall index is calculated using this equation: Rainfall index$_t$ = Rainfall$_t$ – Mean/Standard deviation

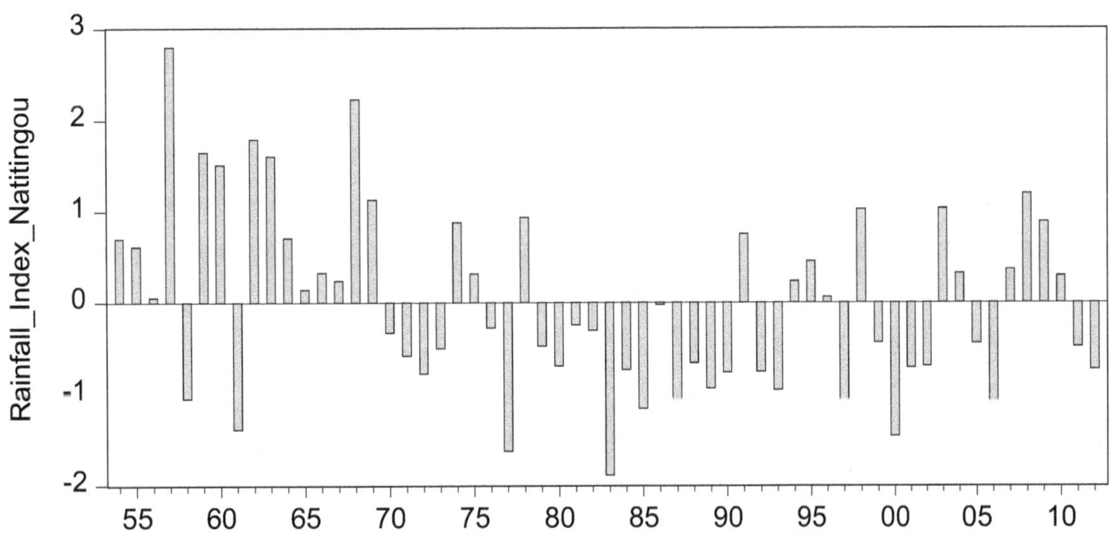

Fig. 8.2 Rainfall index evolution between 1954 and 2012 in Natitingou

5.39, respectively. Even in 1998, Kota Monongou was the most vulnerable and Sirikou the less vulnerable to climate shocks. Indeed, Sirikou has the highest adaptive capacity in 1998 and in 2012. Kota Monongou and Moupémou were the most exposed villages to climate shocks in 1998, while in 2012, the exposure level was similar for all the villages. The villages are classified in terms of vulnerability for the two periods. In 1998, 42.86% of the communities were vulnerable (without accounting for sensitivity), while 53.57% were vulnerable in 2012. In terms of overall vulnerability to climate shocks, 46.43% of the communities were vulnerable in 2012. Among the communities that were vulnerable in 1998, 83.33% were still vulnerable in 2012.

As for adaptive capacity, 50% of the communities lacked it in 1998 and 2012, out of the 14 villages tracked. When considering all the 28 villages in 2012, 53.57% lacked adaptive capacity. The situation of some villages in terms of adaptive capacity has been improved between 1998 and 2012 (e.g., Kandifo Peulh and Tintinmou Peulh), while some have seen their situation worsened (e.g., Gantieco and Kota Monongou). Therefore, at least half of the communities in the basin appear to be not resilient to climate shocks. The situation differs across the five capitals. Lack in financial capital is relatively common among surveyed communities: 50% and 53.57% of the villages lacked financial capital in 1998 and 2012, respectively. Regarding physical, institutional capital, and technology, 57.14% of the surveyed villages lacked it in 1998, while a decrease in this percentage is noted in 2012 (46.43%). Therefore, this capital has been improved among the surveyed communities in the basin during the two periods. Although 35.71% of the communities lacked human capital in 1998, the situation has worsened in 2012; this percentage amounted to 53.57% in 2012. On average, the situation in terms of natural capital has been improved among surveyed communities, although it needs improvements. In 1998, 64.29% lacked natural capital, while in 2012 they were 53.57% lacking this capital. In 1998, on average, social capital level was quite low, as reflected by the 71.43% of surveyed villages lacking this capital. Social capital level has been improved in 2012 compared with the situation in 1998, although a large number of communities still lacked this capital (60.71%). Overall, resilience level is low in the basin.

The degree of vulnerability of the communities across AEZs is also investigated. The communities of AEZ II were the least vulnerable to climate shocks, followed by AEZs I, III, and IV in 2012. Indeed, communities of AEZ IV were the most exposed and the most sensitive to climate shocks, and also had the lowest adaptive capacity. Moreover, communities of AEZ IV had the lowest social capital, whereas communities of AEZ II have the highest. This means that farmers in communities of AEZ II helped one another in mitigating the effects of climate shocks, and this leads to their highest resilience level. Even in 1998, communities of AEZ II had the highest adaptive capacity and were the less exposed to climate shocks, whereas communities of AEZ IV had the lowest adaptive capacity and were the most exposed to these shocks. The highest adaptive capacity level of communities of AEZ II was due to their higher financial capital, and physical, institutional capital and technology in 1998. The social capital of communities of AEZ II has been improved over time. The analyses show that, in 1998, communities of AEZ IV lacked all kind of capital; the situation is alarming for financial, physical, and institutional capital, and technology.

8.5.3 Sensitivity and Uncertainty Analyses

Vulnerability indices for the two periods were computed 1000 times to map their probability distribution. For each dimension of vulnerability, random values were generated between its minimum and maximum values. The generated vulnerability indices for the two periods follow the normal distribution (Figs. 8.3 and 8.4). Moreover, the reliability of the original calculated vulnerability indices is estimated through determination of the range of the standard deviations around the means, and Student's t-tests revealed that they lie within the respective range ($p < 0.05$). As for sensitivity, the values of some indicators have been changed or some indicators were simply disregarded to explore the influence on vulnerability indices. These analyses showed that the indices are sensitive to these changes.

8.5.4 Econometric Results

Panel specification tests have been run through Fisher test and Breusch and Pagan Lagrangian multiplier test. The Fisher test indicates that there are significant individual (village level) effects, implying that pooled ordinary least squared (OLS) would be inappropriate (Prob > F = 0.0.3). The Breusch and Pagan Lagrangian multiplier test indicates the presence of random effects (Prob > chibar^2 = 0.06). Thus, both of these two specification tests

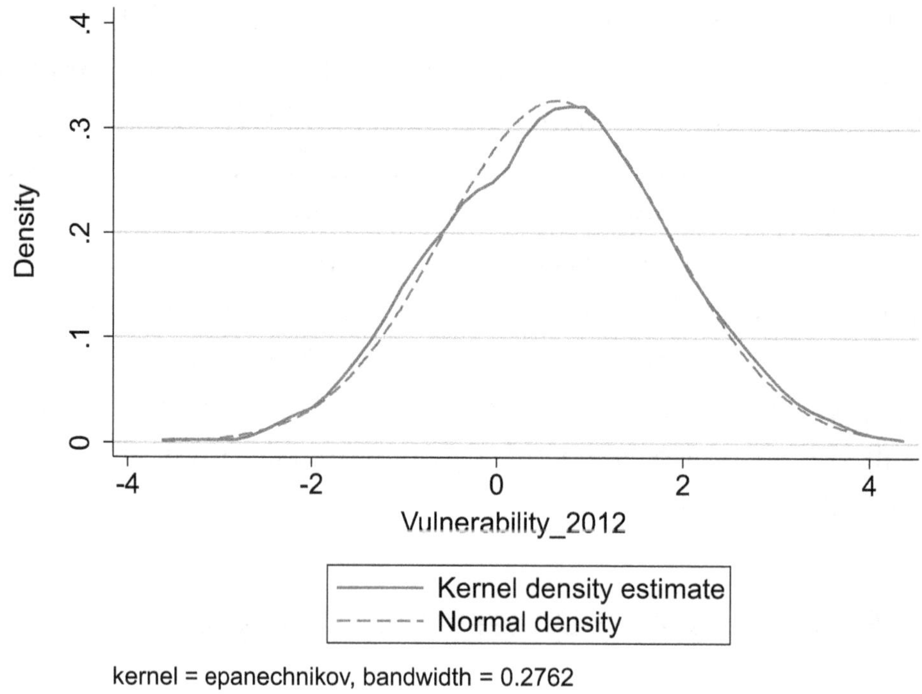

Fig. 8.3 Kernel density of the generated vulnerability index (2012)

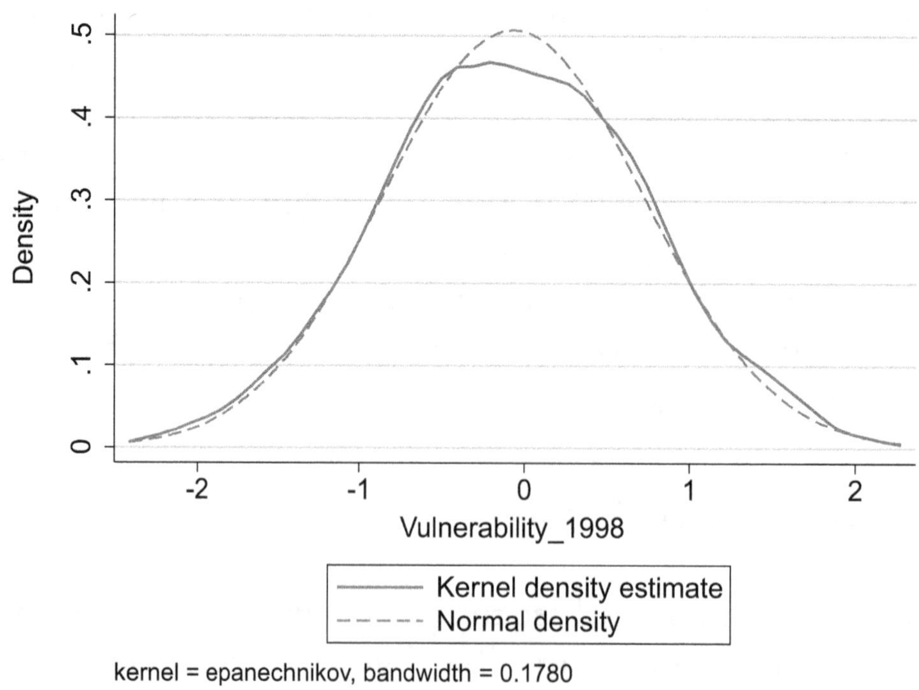

Fig. 8.4 Kernel density of the generated vulnerability index (1998)

indicate that pooled OLS would be inappropriate (at 5% and 10% significance level for the fixed effects and random effects, respectively). The Hausman specification test is used to choose among fixed effects and random effects. Prob > $chi^2 = 0.11$ leads to the strong non-rejection of the null hypothesis that the difference in coefficients is not systematic. As the time period is short (two periods), random effects seem more appropriate than fixed effects. Consequently, the estimation procedure accounts for the likely endogeneity of land tenure in the random effects model. Table 8.4 presents the results of the estimations.

The impact of the two climate variables is non-linear. A given change in temperature from long-term mean will lessen vulnerability to climate shocks up to 0.35 degree Celsius, and beyond this threshold, the impact will be positive, ceteris paribus. This could be explained by the fact that the crops will gain from carbon fertilization under a change less than 0.35 degree Celsius, ceteris paribus. Regarding precipitations, a given change in rainfall from long-term mean will lessen vulnerability to climate shocks up to 10.79%, and beyond this threshold, the impact will be positive, ceteris paribus. However, the coefficients associated to the change in percentage in rainfall from long-term mean and its square are not significant.

Land tenure security leads to strengthening vulnerability to climate shocks, with the impact being non-significant. This finding suggests that farmers within the communities may not be yet taken advantage of their land tenure status in terms of investments in appropriate technologies relative to farming. This chapter considers also off-farm activities in the analyses of vulnerability, therefore the findings need to be analyzed with respect to that. Lack of tenure security may push farmers within the communities to look for off-farm activities, which are less climate dependent, and therefore, they appear to be less vulnerable through diversifying income sources. However, the result needs to be taken with caution, and needs further investigation in terms of socio-cultural elements that may impede the advantage of having secure tenure.

Membership to farmers' labor-sharing groups and to farmers' organizations appears to be useful for farmers. Indeed, through these organizations, farmers receive relevant information regarding how to deal with farming, such as fertilizer requirements, when to use fertilizers and pesticides, and climatic information. However, the impact of membership to farmers' organizations is higher than the impact of membership to farmers' labor-sharing

Table 8.4 Regression results of vulnerability

Dependent variable: vulnerability index			
Variables	Coefficients	Standard errors	*Z*-statistics
Change in rainfall from long-term mean	0.026	0.021	1.26
Square of change in rainfall from long-term mean	−0.001	0.001	−1.64
Change in temperature from long-term mean	37.605**	19.118	1.97
Square of change in temperature from long-term mean	−54.043*	30.805	−1.75
Proportion of households that belong to farmers' labor-sharing groups	1.012*	0.562	1.80
Proportion of households that belong to farmers' organizations	2.776***	0.745	3.73
Density of primary schools within the community	96.583*	55.914	1.73
Percentage of households that have access to electricity	−0.011	0.921	−0.01
Land tenure security	−0.065	0.061	−1.07
Constant	2.012	4.354	0.46
R-squared	Overall = 0.522	Within = 0.536	Between = 0.485

Note: ***, **, * Significant at the 1%, 5%, and 10% levels, respectively. Lower values of the dependent variable (vulnerability) indicate improvement in vulnerability

groups. This could be explained by the fact that farmers had access to extension services mainly through farmers' organizations. These findings confirm the usefulness of social capital in improving welfare as suggested by the literature (Ostrom and Ahn 2007; Ostrom 1994). More primary schools will lessen vulnerability to climate shocks. Indeed, the more the populations are literate, the more they will be able to obtain appropriate information regarding adapting to climate shocks. The impact of the percentage of households that have access to electricity is negative. This finding means that access to electricity is costly to the communities, and it reduces the share of financial means that is invested in adaptation strategies. However, the percentage of households that have access to electricity does not have a significant impact on vulnerability level, ceteris paribus.

8.6 Conclusion

This chapter assesses the vulnerability of communities to climate shocks in the Niger basin of Benin (14 villages tracked between 1998 and 2012, and 14 additional ones in 2012), and analyzes the extent to which land tenure affects vulnerability. First, indices were built for each dimension of vulnerability: adaptive capacity (financial, human, natural, social, physical, and institutional capital, and technology), sensitivity, and exposure. Second, overall vulnerability indices were built. The findings reveal that between 1998 and 2012, the situation of the villages has been improved except for five villages (Kossou, Kpbébéra, Gantiéco, Kota Monongou and Moupémou). Sirikou was the less vulnerable community in 1998 and 2012, whereas the most vulnerable was Kota Monongou. Half of the communities tracked lacked adaptive capacity (through which resilience was analyzed) during the two periods. In 2012, 53.57% of the 28 communities appeared to have a lack in adaptive capacity. Thus, resilience level is low in the basin. On average, communities of AEZ II were the less vulnerable to climate shocks, followed by AEZs I, III, and IV in 2012. The econometric results suggest that farmers' labor-sharing groups, farmers' organizations, and access to primary education have the potential to lessen vulnerability to climate shocks. Tenure security appears to lead to strengthening non-significantly vulnerability to climate shocks.

The situation of the communities in terms of food and nutrition security will be affected if any action is taken, as the farm households are mainly subsistence farmers. Indeed, this could lead to vulnerability to food insecurity. Therefore, public policies should encourage formal and informal social networks that enable group discussions and better information flows and improve adaptation to climate shocks. They should promote access to primary education and raise the awareness of the farmers within the communities on investment in relevant technology and environmental management practices which have the potential to lessen their vulnerability and strengthen their resilience to climate shocks. Moreover, diversification of income sources off the farm can be promoted. Furthermore, they should think about providing timely climate information to the communities. Results indicating differences among villages and AEZs suggest that adaptation technologies should be targeted to the various villages and AEZs to enhance their specific adaptation potential.

Appendix

Table 8.5 Descriptive statistics for indicators used to compute vulnerability in 1998

Indicators	Minimum	Maximum	Mean	Standard deviation
Change in percentage in rainfall from the long-term mean	0.73	38.70	31.43	11.02
Change in degree in temperature from the long-term mean	0.03	0.63	0.20	0.28
Fertilizer-use value per household	1889	306,702	96,438	84,375.77
Herbicide-use value per household	1778	238,944	63,711.69	69,517.54
Yearly income from agricultural off-farm activities per household	0	86,556	11,063.49	22,909.61
Yearly income from non-agricultural off-farm activities per household	0	341,000	96,856.92	85,931.84
Yearly income from cropping per household	101,089	1,817,489	636,540.89	520,734.99
Yearly income from livestock per household	2633	1,116,333	248,289.88	312,268.03
Percentage of households that use plow	0	1	0.61	0.39
Livestock value per household	24,400	2,570,000	1,207,054.11	839,804.36
Amount of credit obtained per household	0	65,556	14,952.38	18,577.30
Number of times of access to extension services per household	0	2	0.71	0.80
Distance from dwelling to food market per household	0	2	0.65	0.54

(*continued*)

Table 8.5 (continued)

Indicators	Minimum	Maximum	Mean	Standard deviation
Distance from dwelling to paved or tarred road per household	0	82	28.80	27.05
Proportion of households within the communities that have access to electricity	0	0	0.1	0.03
Asset value per household	46,331	386,195	188,969.93	96,385.50
Density of primary schools within the community	0	0.019	0.01	0.01
Density of secondary schools within the community	0	0	0	0
Density of high schools within the community	0	0	0	0
Density of maternities within the community	0	0.003	0.0002	0.001
Density of municipality hospitals within the community	0	0	0	0
Density of district hospitals within the community	0	0.003	0.0002	0.001
Average household head formal education	0	2	0.74	0.82
Bush and valley bottom land-use size per household	1	10	3.88	2.97
Irrigated land-use size per household	0	0	0.01	.02
Proportion of households within the communities that belong to labor-sharing groups	0	1	0.31	0.38
Proportion of households within the communities that belong to farmers' organizations	0	1	0.79	0.33
Amount of financial assistance per household	0	20,000	3178.57	5653.17

Table 8.6 Descriptive statistics for indicators used to compute vulnerability in 2012

Indicators	Minimum	Maximum	Mean	Standard deviation
Change in percentage in rainfall from the long-term mean	−13.19	13.02	−2.88	8.67
Change in degree in temperature from the long-term mean	0.61	0.64	0.62	0.01
Proportion of households within the communities that experienced flood	0	1	0.48	0.30
Proportion of households within the communities that experienced droughts	0	1	0.46	0.27
Proportion of households within the communities that experienced strong winds	0.75	1	0.95	0.07
Proportion of households within the communities that experienced heat waves	0	1	0.59	0.31
Proportion of households within the communities that experienced erratic rainfall	0.40	1	0.87	0.17
Proportion of households within the communities that experienced heavy rainfall	0.37	1	0.80	0.17
Proportion of households within the communities that experienced a change in planting date	0.30	1	0.81	0.21
Proportion of households within the communities that report a decrease in yield	0.15	0.95	0.55	0.25
Fertilizer-use value per household	3800	273,600	100,208.77	76,732.74
Herbicide-use value per household	0	121,762.50	31,990.01	29,846.48
Insecticide-use value per household	0	101,400	24,465.26	27,653.01
Yearly income from agricultural off-farm activities per household	0	275,100	30,596.35	55,475.13
Yearly income from non-agricultural off-farm activities per household	42,300	1,229,289	293,713.73	232,765.30
Yearly income from cropping per household	333,925	2,902,234.19	1,423,760.69	559,075.06
Yearly income from livestock per household	1166.67	529,925	78,372.93	104,789.54

Proportion of households within the communities that use plow	0	1	0.54	0.39
Livestock value per household	60,872.22	9,165,411.82	1,194,416.98	1,814,838.78
Amount of credit obtained per household	0	132,500	19,357.33	27,047.84
Number of times of access to extension services per household	0	4.65	1.18	1.13
Distance from dwelling to food market per household	0.01	1.15	0.44	0.26
Distance from dwelling to paved or tarred road per household	0.12	53.15	10.85	14.11
Proportion of households within the communities that have access to electricity	0	0.75	0.23	0.21
Asset value per household	78,345	597,968.75	309,607.40	142,475.38
Density of primary schools within the community	0	0.02	0.005	0.004
Density of secondary schools within the community	0	0.003	0.0004	0.001
Density of high schools within the community	0	0.003	0.0001	0.001
Density of maternities within the community	0	0.003	0.001	0.001
Density of municipality hospitals within the community	0	0.003	0.0002	0.0007
Density of district hospitals within the community	0	0.003	0.0006	0.001
Average household formal education	0.05	3.60	1.71	0.93
Bush and valley bottom land-use size per household	0.52	10.98	5.29	2.76
Irrigated land-use size per household	0	2.00	0.10	0.38
Proportion of households within the communities that belong to labor-sharing groups	0	0.85	0.24	0.22
Proportion of households within the communities that belong to farmers' organizations	0	0.95	0.36	0.30
Amount of financial assistance per household	0	25,000	3364.58	6375.80
Value of assistance in nature per household	0	6002.99	1902.95	1810.10

Table 8.7 Factor scores for financial capital

	Components		
	1	2	3
Fertilizer-use value per household	0.37	0.01	−0.04
Herbicide-use value per household	0.37	−0.09	−0.04
Insecticide-use value per household	0.33	0.01	0.00
Yearly income from agricultural off-farm activities per household	0.09	−0.61	0.07
Yearly income from non-agricultural off-farm activities per household	−0.07	0.21	−0.65
Yearly income from cropping per household	0.03	0.53	−0.01
Yearly income from livestock per household	−0.14	0.15	0.65
% of variance	36.30	20.92	17.29

Table 8.8 Factor scores for physical and institutional capital, and technology

	Components				
	1	2	3	4	5
Proportion of households within the communities that use plow	0.00	0.10	0.31	−0.14	0.25
Livestock value per household	−0.04	0.13	0.40	0.17	0.00
Amount of credit obtained per household	−0.05	0.09	−0.05	0.00	0.71
Frequency of access to extension services per household	0.00	0.10	0.42	−0.13	−0.21
Distance from dwelling to food market per household	−0.11	0.42	0.20	−0.12	−0.08
Distance from dwelling to paved or tarred road per household	−0.24	0.12	0.00	−0.10	−0.20
Proportion of households within the communities that have access to electricity	0.03	0.20	−0.15	−0.34	0.28
Household asset value per household	0.26	−0.22	0.14	0.10	0.12
Density of primary schools within the community	−0.07	0.03	−0.01	0.66	0.03
Density of secondary schools within the community	0.29	−0.04	−0.01	0.02	−0.15
Density of high schools within the community	−0.09	0.38	0.07	0.19	0.09
Density of maternities within the community	0.33	−0.05	0.02	−0.17	−0.01
Density of municipality hospitals within the community	−0.02	0.31	−0.02	0.12	0.10
Density of district hospitals within the community	0.35	−0.08	−0.01	−0.08	−0.12
% of variance	20.74	18.46	15.82	10.16	9.46

Table 8.9 Factor scores for human, natural, and social capital

	Components		
	1	2	3
Average household head formal education	−0.123	−0.295	0.168
Bush and valley bottom land-use size per household	−0.01	0.263	0.483
Irrigated land-use size per household	0.07	0.082	−0.711
Proportion of households within the communities that belong to labor-sharing groups	0.413	−0.196	0.034
Proportion of households within the communities that belong to farmers' organizations	0.344	−0.057	0.118
Amount of financial assistance per household	−0.025	0.269	0.067
Value of assistance in nature per household	−0.257	0.72	−0.041
% of variance	29.756	19.296	15.404

Table 8.10 Factor scores for exposure and sensitivity

	Components		
	1	2	3
Change in temperature from the long-term mean	−0.124	0.486	0.063
Proportion of households that experienced flood over the last 20 years	0.265	−0.189	−0.227
Proportion of households that experienced droughts over the last 20 years	0.201	0.137	−0.02
Proportion of households that experienced strong winds over the last 20 years	0.218	0.051	0.037
Proportion of households that experienced heat waves over the last 20 years	0.221	−0.141	0.012
Proportion of households that experienced erratic rainfall over the last 20 years	0.019	0.497	−0.083
Proportion of households that experienced heavy rainfall over the last 20 years	0.238	−0.105	0.174
Proportion of households that experienced a change in planting date over the last 20 years	−0.127	−0.167	0.023
Proportion of households that experienced a decrease in yield over the last 20 years	0.039	0.11	−0.54
Change in rainfall from the long-term mean	0.023	0.109	0.502
% of variance	35.474	16.798	15.752

Table 8.11 Descriptive statistics of the variables used in the regression

Variables	Mean	Standard deviation	Minimum	Maximum
Vulnerability	0.0002	1.295	−2.48	3.14
Change in rainfall from long-term mean	8.554	18.868	−13.19	38.7
Square of change in rainfall from long-term mean	420.7	560.743	0.533	1497.69
Change in temperature from long-term mean	0.479	0.254	0.03	0.64
Square of change in temperature from long-term mean	0.292	0.165	0.001	0.410
Proportion of households that belong to farmers' labor-sharing groups	0.256	0.319	0	1
Proportion of households that belong to farmers' organizations	0.480	0.398	0	1
Density of primary schools within the community	0.006	0.005	0	0.019
Percentage of households that have access to electricity	0.153	0.206	0	0.75
Land tenure security	92.826	12.209	40.754	100

Notes

1. The AfDB through its Ten-Year strategy (called the 'High 5s'), is committed to improving food security and rural livelihoods by tackling the most important constraints on agricultural productivity, and to building resilience to climate change (AfDB 2016).
2. Yegbemey et al. (2013) distinguished between inheritance, gifting, renting, and purchasing in Northern Benin.

References

Adger, N. W. (2006). Vulnerability. *Global Environmental Change, 16*(3), 268–281.

AfDB. (2016). *Annual Development Effectiveness Review 2016: Accelerating the Pace of Change*. Abidjan: African Development Bank Group.

AfDB, OECD, & UNDP. (2015). *African Economic Outlook 2015: Regional Development and Spatial Inclusion*. African Development Bank, Organisation for Economic Co-operation and Development, United nations Development.

Baltagi, B. H. (2008). *Econometric Analysis of Panel Data* (4th ed.). West Sussex: John Wiley & Sons.

Béné, C., Wood, R. G., Newsham, A., & Davies, M. (2012). *Resilience: New Utopia or New Tyranny? Reflection About the Potentials and Limits of the Concept of Resilience in Relation to Vulnerability Reduction Programs.* Institute of Development Studies (IDS), Working Paper 2012(405) and Center for Social Protection (CSP), Working Paper 006, Brighton.

Berman, R., Quinn, C., & Paavola, J. (2012). The Role of Institutions in the Transformation of Coping Capacity to Sustainable Adaptive Capacity. *Environmental Development, 2*, 86–100.

Brooks, N., Adger, N. W., & Kelly, M. (2005). The Determinants of Vulnerability and Adaptive Capacity at the National Level and Implications for Adaptation. *Global Environmental Change, 15*(2), 151–163.

Callo-Concha, D., Gaiser, T., Webber, H., Tischbein, B., Müller, M., & Ewert, F. (2013). Farming in the West African Sudan Savanna: Insights in the Context of Climate Change. *African Journal of Agricultural Research, 8*(38), 4693–4705.

Chagutah, T. (2013). Land Tenure Insecurity, Vulnerability to Climate-Induced Disaster and Opportunities for Redress in Southern Africa. *Jàmbá: Journal of Disaster Risk Studies, 5*(2), 1–8.

Cutter, S. L., Barnes, L., Berry, M., Burton, C., Evans, E., Tate, E., & Jennifer, W. (2008). A Place-Based Model for Understanding Community Resilience to Natural Disasters. *Global Environmental Change, 18*(4), 598–606. https://doi.org/10.1016/j.gloenvcha.2008.07.013.

Davies, M., Guenther, B., Leavy, J., Mitchell, T., & Tanner, T. (2008). 'Adaptive Social Protection': Synergies for Poverty Reduction. *IDS Bulletin, 39*(4), 105–112.

Davies, M., Oswald, K., & Mitchell, T. (2009). Climate Change Adaptation, Disaster Risk Reduction and Social Protection. In OECD, *Promoting Pro-Poor Growth: Social Protection* (pp. 201–217). Paris: Organisation for Economic Cooperation and Development

Deressa, T. T., Hassan, R. M., & Ringler, C. (2009). *Assessing Household Vulnerability to Climate Change: The Case of Farmers in the Nile Basin of Ethiopia.* International Food Policy Research Institute, Discussion Paper 00935, Washington, DC.

Deressa, T. T., Hassan, R. M., & Ringler, C. (2008). *Measuring Ethiopian Farmers' Vulnerability to Climate Change Across Regional States.* International Food Policy Research Institute, Discussion Paper 00806, Washington, DC.

Devereux, S., & Sabates-Wheeler, R. (2004). *Transformative Social Protection.* Working Paper 232, Institute of Development Studies, Brighton.

Dixon, R. K., Smith, J., & Guill, S. (2003). Life on the Edge: Vulnerability and Adaptation of African Ecosystems to Global Climate Change. *Mitigation and Adaptation Strategies for Global Change, 8*(2), 93–113.

Dunford, R., Harrison, P. A., Jäger, J., Rounsevell, M. D., & Tinch, R. (2015). Exploring Climate Change Vulnerability Across Sectors and Scenarios Using Indicators of Impacts and Coping Capacity. *Climatic Change, 128*(3), 339–354.

Economic Commission for Africa. (2003). Land Tenure Systems and Sustainable Development in Southern Africa. ECA/SA/EGM. Land/2003/2, United Nations Economic Commission for Africa, Southern Africa Office, Lusaka.

Etwire, P. M., Al-Hassan, R. M., Kuwornu, J. K., & Osei-Owusu, Y. (2013). Application of Livelihood Vulnerability Index in Assessing Vulnerability to Climate Change and Variability in Northern Ghana. *Journal of Environment and Earth Science, 3*(2), 157–170.

Fofana, I. (2011). *Simulating the Impact of Climate Change and Adaptation Strategies on Farm Productivity and Income: A Bioeconomic Analysis.* Discussion Paper 01095, International Food Policy Research Institute, Washington, DC.

Füssel, H.-M. (2007). Vulnerability: A Generally Applicable Conceptual Framework for Climate Change Research. *Global Environmental Change, 17*(2), 155–167.

Gavian, S., & Fafchamps, M. (1996). Land Tenure and Allocative Efficiency in Niger. *American Journal of Agricultural Economics, 78*(2), 460–471.

Holling, C. S. (1973). Resilience and Stability of Ecological Systems. *Annual Review of Ecology and Systematics, 4*(1973), 1–23.

IFPRI, & LARES. (1998). *National Survey of Small Farmers in Benin.* International Food Policy Research Institute, and Laboratoire d'Analyse Régionale et d'Expertise Sociale, Washington, DC.

IPCC. (2013). *Climate Change 2013: The Physical Science Basis. Working Group I Contribution to the Fifth Assessment Report of the Intergovernmental Panel on Climate Change*. Cambridge: Cambridge University Press.

IPCC. (2014). *Climate Change 2014: Impacts, Adaptation, and Vulnerability. Part B: Regional Aspects. Contribution of Working Group II to the Fifth Assessment Report of the Intergovernmental Panel on Climate Change* (edited by Barros, V. R., Field, C. B., Dokken, D. J., Mastrandrea, M. D., Mach, K. J., Bilir, T. E., Chatterjee, M., et al.). Cambridge: Cambridge University Press.

Islam, M. M., Sallu, S., Hubacek, K., & Paavola, J. (2014). Vulnerability of Fishery-Based Livelihoods to the Impacts of Climate Variability and Change:

Insights from Coastal Bangladesh. *Regional Environmental Change, 14*(1), 281–294.

Jayne, T. S., Yamano, T., Weber, T. M., Tschirley, D., Benfica, R., Chapoto, A., & Zulu, B. (2003). Smallholder Income and Land Distribution in Africa: Implications for Poverty Reduction Strategies. *Food Policy, 28*, 253–275.

Kelly, M., & Adger, W. N. (2000). Theory and Practice in Assessing Vulnerability to Climate Change and Facilitating Adaptation. *Climatic Change, 47*(4), 325–352.

Klein, R. J., Nicholls, R. J., & Thomalla, F. (2003). Resilience to Natural Hazards: How Useful Is This Concept? *Environmental Hazards, 5*(1–2), 35–45.

Lokonon, B. O. K. (2017). *Farmers' Vulnerability to Climate Shocks: Insights from the Niger Basin of Benin.* Working Paper Series No. 248, African Development Bank, Abidjan.

Mahadevia, D. (2011). *Tenure Security and Urban Social Protection in India.* Research Report 05, Institute of Development Studies, Centre for Social Protection, Sussex.

Malone, E. L. (2009). *Vulnerability and Resilience in the Face of Climate Change: Current Research and Needs for Population Information.* PNWD-4087, Battelle Memorial Institute, Washington.

Metropolis, N., & Ulam, S. (1949). The Monte Carlo Method. *Journal of the American Statistical Association, 44*(247), 335–341.

Ostrom, E. (1994). Constituting Social Capital and Collective Action. *Journal of Theoretical Politics, 6*(4), 527–562.

Ostrom, E., & Ahn, T. K. (2007). *The Meaning of Social Capital and Its Link to Collective Action.* Bloomington: Indiana University, Workshop in Political and Policy Analysis.

Pearson, K. (1901). On Lines and Planes of Closest Fit to Systems of Points in Space. *Philosophical Magazine, 2*, 559–572.

République du Bénin. (2014). *Analyse Globale de la Vulnérabilité et de la Sécurité Alimentaire.* Cotonou, Benin: INSAE.

Schickele, R. (1952). Theories Concerning Land Tenure. *Journal of Farm Economics, 34*(5), 734–744.

Scoones, I. (1998). *Sustainable Rural Livelihoods: A Framework for Analysis.* Institute of Development Studies, Working Paper 72, Brighton.

Shewmake, S. (2008). *Vulnerability and the Impact of Climate Change in South Africa's Limpopo River Basin.* International Food Policy Research Institute, Discussion Paper 00804, Washington, DC.

Simane, B., Zaitchik, B. F., & Foltz, J. D. (2016). Agroecosystem Specific Climate Vulnerability Analysis: Application to the Livelihood Vulnerability

Index to a Tropical Highland Region. *Mitigation and Adaptation Strategies for Global Change, 21*(1), 39–65.

Solórzano, A. (2016). *Can Social Protection Increase Resilience to Climate Change?: A Case Study of Oportunidades in Rural Yucatan, Mexico.* IDS Working Paper 465, Center for Social Protection, Institute of Development Studies, Brighton.

Stern, N. (2008). *Key Elements of a Global Deal on Climate Change.* London: London School of Economics and Political Science.

Tesso, G., Emana, B., & Ketema, M. (2012). Analysis of Vulnerability and Resilience to Climate Change Induced Shocks in North Shewa, Ethiopia. *Agricultural Sciences, 3*(6), 871–888.

Turner, B. L. (2010). Vulnerability and Resilience: Coalescing or Paralleling Approaches for Sustainability Science? *Global Environmental Change, 20*(4), 570–576. https://doi.org/10.1016/j.gloenvcha.2010.07.003.

Vincent, K. (2007). Uncertainty in Adaptive Capacity and the Importance of Scale. *Global Environmental Change, 17*(1), 12–24.

Vincent, K., & Cull, T. (2014). Using Indicators to Assess Climate Change Vulnerabilities: Are There Lessons to Learn for Emerging Loss and Damage Debates. *Geography Compass, 8*(1), 1–12.

Yegbemey, R. N., Yabi, A. J., Tovignan, D. S., Gantoli, G., & Kokoye, H. S. (2013). Farmers' Decisions to Adapt to Climate Change Under Various Property Rights: A Case Study of Maize Farming in Northern Benin (West Africa). *Land Use Policy, 34*, 168–175. https://doi.org/10.1016/j.landusepol.2013.03.001.

9

The Impact of Agricultural Insurance on the Demand for Supplemental Irrigation: A Randomized Controlled Trial Experimental Evidence in Northern Ghana

Francis Hypolite Kemeze

9.1 Introduction

The vast majority of smallholder farmers in sub-Saharan Africa (SSA) are dependent on rainfed agriculture for their livelihoods, and they are more often afflicted by the vagaries of drought risk (Elagib 2014; Shiferaw et al. 2014). In fact, rainfed agriculture provides about 95 percent of SSA's food and feed (FAO 2007) and it is the principal source of livelihood for more than 70 percent of the population (Hellmuth et al. 2007). Therefore, for millions of poor smallholder farmers, drought poses a major challenge that can critically restrict options, limit development and pull farmers into poverty trap.

Given the underlined threats of drought on smallholder farmers' livelihoods in SSA, drought preparedness and adaptation become a key priority for any policy intended to help smallholder farmers. In developed countries,

F. H. Kemeze (✉)
International Food Policy Research Institute, Accra, Ghana

A. Shimeles et al. (eds.), *Building a Resilient and Sustainable Agriculture in Sub-Saharan Africa*, https://doi.org/10.1007/978-3-319-76222-7_9

risk-transfer approaches such as insurance have played a role in mitigating drought risk but they have generally not been available in developing countries where insurance markets are limited and are not oriented toward the poor. Recent advances in climate science help in the development of a new type of insurance called weather index insurance[1] that offers new opportunities for managing drought risk in areas where such services were difficult to deliver due to high transaction costs related to poor infrastructure and the classical adverse selection and moral hazard problems in providing financial services. Index insurance is a type of insurance that is linked to an index, such as rainfall, temperature, humidity or crop yields rather than actual loss which is difficult to observe. Access to these risk-transfer services can help protect poor farmers against climate variability while promoting the uptake of productivity enhancing technologies.

While the potential benefits of index insurance are great, its implementation can be difficult (Miranda 1991). The results of most index insurance pilot programs however have been disappointed, with the demand disappearing as soon as the subsidy is eliminated (Farrin and Miranda 2015). Also, because of the more pronounced infrastructural and technology gaps in developing countries, there is the disadvantage that the payoff of the weather derivative does not perfectly correlate to the actual shortfall in the underlying exposure. This is the so-called basis risk. Basis risk refers to the potential mismatch between the index trigger and actual on-farm losses. Besides that, the true benefit of index insurance at the smallholder farmers' level is very puzzled as the insurance does not replace the crop loss. And because of the systemic nature of the event, when it occurs it affects the whole community, the local market included. So, price of staple food goes up and that reduces the value of the insurance pay out and reduces the ability of smallholder farmers to smooth their consumption. Index insurance policies rarely issue indemnity payments due to high deductibles and low-coverage levels.

Investment in water management in rainfed agriculture is another side of novel drought adaptation strategy, particularly in SSA where rainfed agriculture plays such an important economic role. Supplemental irrigation (SI) is one possible water management investment that can help overcoming the challenge of water deficit of rainfed crops in semiarid areas (Rockström et al. 2010).

SI is defined as the application of additional water to otherwise rainfed crops, when rainfall fails to provide essential moisture for normal plant growth, to improve and stabilize productivity (Fox and Rockström 2000; Oweis and Hachum 2006, 2012). SI is a simple but highly effective technology that allows farmers to plant and manage crops at the optimal time, without being at the mercy of unpredictable rainfall. All sources of water can be used for SI systems, including runoff harvested water, surface water, underground water, treated industrial waste water. SI contributes to smallholder farmers' livelihoods in three ways: (1) improves yield, (2) stabilizes production from year to year, and (3) provides suitable conditions for economic use of higher technology inputs. The critical importance of SI lies in its capacity to bridge dry spells and thereby reduce risks of drought in rainfed agriculture in SSA. By reducing risk, SI provides smallholder farmers with the necessary incentive for investments in improved production technologies.

Despite the underlying contributions toward farmers' livelihood, SI is still a rare innovation among smallholder farmers in SSA.

While both drought index insurance and SI address the risk of drought, they do so in very different fashions. As such, smallholder farmers potentially view drought index insurance and SI as either substitutable or complementary drought risk management instruments, depending on various factors such as farmers' experience with drought, whether the risks are related to crop failure or to the additional costs of SI during dry spells or the structure of the insurance contract.

This chapter makes use of the randomized controlled trials experiment to shed light on the existing debate whether drought index insurance and SI as two novel drought risk management instruments are substitute or complementary.

9.2 Weather Index Insurance and Supplemental Irrigation: Previous Studies

SI appears to offer more benefit to farmers than drought index insurance. However, in the recent literature in developing countries, attention have been majoritarily directed toward drought index insurance. The question is why is it so? Do farmers prefer most drought index insurance compared to SI? The literature does not adequately answer this question. Very few studies have looked at the interaction between SI and index insurance. Studies that jointly analyzed SI and weather index insurance are Foudi and Erdlenbruch (2012) in France, Buchholz and Musshoff (2014) in Germany, Barham et al. (2011), Dalton et al. (2004), Lin et al. (2008), and Mafoua and Turvey (2003) in the USA.

Foudi and Erdlenbruch (2012) in analyzing the way French farmers manage drought risk found that SI serves as a self-insurance to farmers. They further found that a farmer's decision whether to irrigate (or not) depends on his decision to purchase insurance (or not). Insurance decreases the probability of adopting irrigation. Thus, the offered yield insurance, as they further conclude, may serve to decrease the amount of water used for irrigation.

Buchholz and Musshoff (2014) investigate the potential of index insurance to cope with the economic disadvantages for farmers resulting from a reduction in water quotas and increased water pieces. They do that by comparing crop portfolios without and with index insurance and they found that the use of weather index insurance offsets the loss in the farmer's certainty equivalent resulting from moderate reductions in water quotas and water price increases. They also found that weather index insurance has the potential to substantially alter farm plans and the optimal irrigation water demand. Barham et al. (2011) compare discrete combinations of multiple-peril crop insurance and varying levels of irrigation in a stochastic simulation setting for a cotton farm in Texas. Their findings show that crop insurance is particularly beneficial at lower irrigation levels.

Dalton et al. (2004) evaluate the benefits of multiple-peril crop insurance and the investment in SI for potato production in Maine. Using a biophysical simulation model, the authors derived the risk management benefits of SI and crop insurance over nonirrigated uninsured production. The authors found that crop insurance programs are inefficient at reducing producer exposure to weather-related production risk in humid regions. They also found the risk management benefits from SI to be scalable and technology dependent. Increasing the scale of technology adoption increases the risk management benefits of SI. Lin et al. (2008) investigate irrigation strategies for maize production in Georgia in case of varying water prices and the availability of a precipitation-based weather derivative. Their results reveal that the derivative performs relatively poorly in terms of increasing the estimated certainty equivalent revenues and has no impact on the amount of irrigation water used.

Mafoua and Turvey (2003) provide a conceptual regression model using annual cross-sectional data from New Jersey. They demonstrate that precipitation-based weather derivatives may enable farmers to hedge against irrigation costs in drought years.

The literature does not make it clear whether drought index insurance and SI are substitutes or complementarity risk management tools. The first line of research shows that SI and drought index insurance are substitutes in the sense that drought index insurance may be used to reduce the amount of water used. The second line of research demonstrates the benefit of SI over drought index insurance as drought index insurance is inefficient at reducing farmers' exposure to drought, performs poorly in terms of increasing farmers' revenues and has no impact on the amount of irrigation water used. The last line of literature considered drought index insurance and SI as complementary risk management tools because drought index insurance can be used to offset the high cost of SI during drought years.

It appears therefore worthy to contribute to this interesting debate by assessing the impact of drought index insurance on the demand for SI in developing countries.

9.3 Methodology

9.3.1 Experimental Design

In 2014, with the support of USAID-BASIS fund a randomized control trials (RCT) study was undertaken by the Ohio State University and the African Center for Economic Transformation, in collaboration with the University of Ghana, in order to investigate the impact of drought index insurance on the adoption of agricultural technology including SI among smallholder farmer in Northern Ghana.[2]

Following a list of farmers provided by the Rural Community Banks (RCB), a rural inclusive financial service provider in charge of all the three Northern regions of the study area, and based on a preliminary field visit with this institution and their farmers, a selection of 279 farmer groups out of 791 groups was made. A baseline survey was then conducted in early 2015 in order to gather household's demographic and socioeconomics characteristics necessary to assure similarity among potential assigned treatments and control groups. During the baseline survey an experimental filed survey was also undertaken to elicit smallholder farmers' willingness to pay (WTP) for SI.

For the baseline survey, six farmers were randomly selected from each of the 279 farmer groups using a uniform distribution with the intent to interview the first three farmers and the second three farmers as backup in case the first three farmers were unavailable for the interview. A total of 777 farmers were interviewed. Table 9.1 presents the composition of the sample size by region and gender.

Based on the information collected from the baseline survey, the 279 farmer groups were randomly assigned into three groups: (1) Control: smallholder farmers were offered conventional loans, no drought index insurance; (2) Treatment 1: smallholder farmers were offered insured loans where the farmers themselves were policy holders and any payouts are made directly to them; (3) Treatment 2: smallholder farmers were offered insured loans where bank was the policy holder and payouts were to be made to bank and credited toward the outstanding debt of farmer groups. Randomization took place within two strata; region and loan status of the farmers to ensure balance impact across region and loan status. Table 9.2 presents the preliminary number of farmers within region and treatment. Farmer groups were then invited to apply for loans (control) or insured loans (treatment 1 and 2).

The intervention took place during the 2015 farming season followed by a follow-up survey of the same farmers who were included in the baseline survey and we also repeated the WTP for SI experiment.

Table 9.1 Sample size by region and gender

Region	Male	Female	Total
Northern	156	142	298
Upper West	64	20	84
Upper East	182	213	395
Total	402	375	777

Table 9.2 Preliminary farmers in treatment and control categories by regions [2]

Treatment	Northern	Upper West	Upper East	Total
Control	103	27	131	261
Treatment 1	96	33	132	261
Treatment 2	99	24	132	255
Total	298	84	395	777

9.3.2 Data

We use the demand for SI outcomes to assess the impact of index insurance. The demand for SI and socioeconomic characteristics were collected before and after the intervention.

To elicit the demand for SI, a Contingent Valuation Method (CVM) was employed. Since SI systems are not yet available in Ghana, the CVM method is convenient for this study. Following Arrow et al. (1993) recommendations which lead to maximize the reliability of the CVM, we employed a single-bounded dichotomous choice questions. Designing contingent valuation questions in the form of hypothetical referenda in which respondents are told how much they would have to pay for each product before asking them to respond by a simple yes or no answer was used in this study first, to imitate the real world market where a price is given and the consumer chooses to purchase or not to purchase the product at the stated price. Second, to avoid bias induced by asking follow-up WTP questions as with double bounded dichotomous choice questions (Chantarat et al. 2009). The seasonal subscription charged fee per acre is one of seven values which were determined based on the estimated mean fee per acre of GHC 20.00, charged for similar services in Ghana, Burkina Faso, Kenya and India (Alhassan et al. 2013; Chandrasekaran et al. 2009; Zongo et al. 2015). The seven bid values used in the study include the estimated mean fee with six additional values that are ±5 percent, ±15 percent, and ±25 percent of the estimated means fee (Bids: GHC 15, 17, 19, 20, 21, 23, & 25). Each of this bid was randomly assigned to each respondent.

9.3.3 Sample Size, Power Calculations and Minimum Detectable Effect

Choosing an appropriate sample size in experimental research matter as it increases the probability to detect an effect, assuming there is a genuine effect which is to be detected. This is the so-called power of the experiment. It measures the ability of a test to reject the null hypothesis when it should be rejected. Duflo et al. (2008) define power of an experiment as the probability that, for a given effect size and a given statistical significance level, one can reject the hypothesis of zero effect. The power of the experiment is affected by the sample size, the statistical significant level and other design choices. The minimum accepted level of power is considered to be 80 percent in the literature, which signifies that there is an eight in ten chance of detecting a difference of the specified effect size (Bloom 1995; Duflo et al. 2008). The statistical significant level (p value) is the probability of a type I error (that is the probability we reject the null hypothesis when it is in fact true). Usually 5 percent is used.

The sample size for our experiment is at group level since each of our farmers belongs to a farmer-based organization (and more often are member of the same household). This is to reduce potential spillover effect. We then need to pick our sample size such as to minimize the effect size taken into consideration power 80 percent and statistical significant level 5 percent. For this purpose, it is useful to measure precision in terms of minimum detectable effects (Bloom 1995, 2006). A minimum detectable effect is the smallest true treatment effect that a research design can detect with confidence.

As we intend to test the hypothesis whether drought index insurance has effect on the demand for SI at farmer level, our sample size is at individual level. Our outcome variable is a binary "yes" or "no" WTP. Following Duflo et al. (2008), the minimum detectable effect (MDE) for this binary outcome given power (κ), significance level (α), sample size (N) of identical size (n), and portion of subjects allocated to treatment group (P) is given by:

$$\text{MDE} = \frac{1}{c}\frac{M_{N-2}}{\sqrt{p(1-p)N}}\sqrt{\rho + \frac{1-\rho}{n}\pi(1-\pi)}$$

where $M_{j-2} = t_{\alpha/2} + t_{1-k}$, for a two-sided test;

C is the compliance rate for the treatment; P is the proportion of treated sample; σ is the variance in the outcome variable; ρ is the intra-class correlation of the farmers within each group; π is the proportion of the study population that would have a value of one for the binary variable in the absence of the program (Bloom 1995).

In order to calculate the MDE for our experiment, we use the baseline survey to estimate the variance for the outcome variable. Table 9.3 presents the value of the parameters we used to estimate the MDE. The results of the MDE estimations are presented in Table 9.4.

Table 9.3 Parameter assumptions for MDE calculations

Number of members per group (*n*)	3
Proportion of sample in treatment (*P*)	0.5
$t_{\alpha/2}$	1.65
t_{1-k}	0.84
Power (*k*)	0.8
Significance level (α)	0.05
Share of treatment group actually treated (*C*)	0.5

Table 9.4 MDEs calculations for binary outcome variables

Intra-class correlation	WTP for Canal SIES	WTP for drip SIES
Sample group size 100		
0.05	0.089	0.092
0.2	0.1	0.104
0.4	0.114	0.118
0.6	0.126	0.131
Sample group size 150		
0.05	0.072	0.075
0.2	0.082	0.085
0.4	0.093	0.096
0.6	0.103	0.107
Sample group size 180		
0.05	0.067	0.069
0.2	0.075	0.078
0.4	0.085	0.088
0.6	0.094	0.097
$\pi(1-\pi)$	0.148	0.153

The MDE is sensitive to the sample size. It is lower when the sample size is bigger. We choose our sample size to be 279 roughly equally divided among the groups. That corresponds to an individual total size of 837.

9.3.4 Study Area

This study took place in the Northern Savannah zone of Ghana (Northern, Upper East and Upper West) to assess potential drought risk management tools among smallholder farmers. The choice of the study area is based on its agricultural contribution to the entire country and the great threat of drought that warms agriculture, the main activity in the zone. The Northern Savannah zone is the largest agriculture zone in Ghana. Most of the nation's supply of maize, rice, millet, sorghum, yam, tomatoes, cattle, sheep, goat and cotton are grown in Northern Savannah. This is because the Northern Savannah zone carries two-thirds of the nation's grassland.

The Northern Savannah agro-ecological zone of Ghana is characterized by unimodal rainfall of short duration, high incidence of droughts and excessive evapotranspiration[3] allowing only 4 to 5 months of farming and 7 to 8 months of extended dry season. Yet agriculture in the zone is predominantly rainfed with less than 0.4 percent of the agricultural land irrigated. As a result, droughts often impact severely on livelihoods in the area (Dietz et al. 2004; Laube et al. 2008; Van de Giesen et al. 2010). The effects of drought on food production in the area are greater than anywhere else in the country (MOFA 2007; EPA 2012). Rainfall variability in the zone is exacerbated by climate change, resulting in a rise in the frequency of droughts (EPA 2007; Hesselberg and Yaro 2006). Adaptation policies with regard to drought in this region have, however, been insufficient (EPA 2012; Yaro 2013). Figure 9.1 presents the map of the study area. Figure 9.1 further presents the groupment of farmers and the distance between their household and a nearby river. GPS were recorded and those were used to estimate the distance between the household and the nearby river.

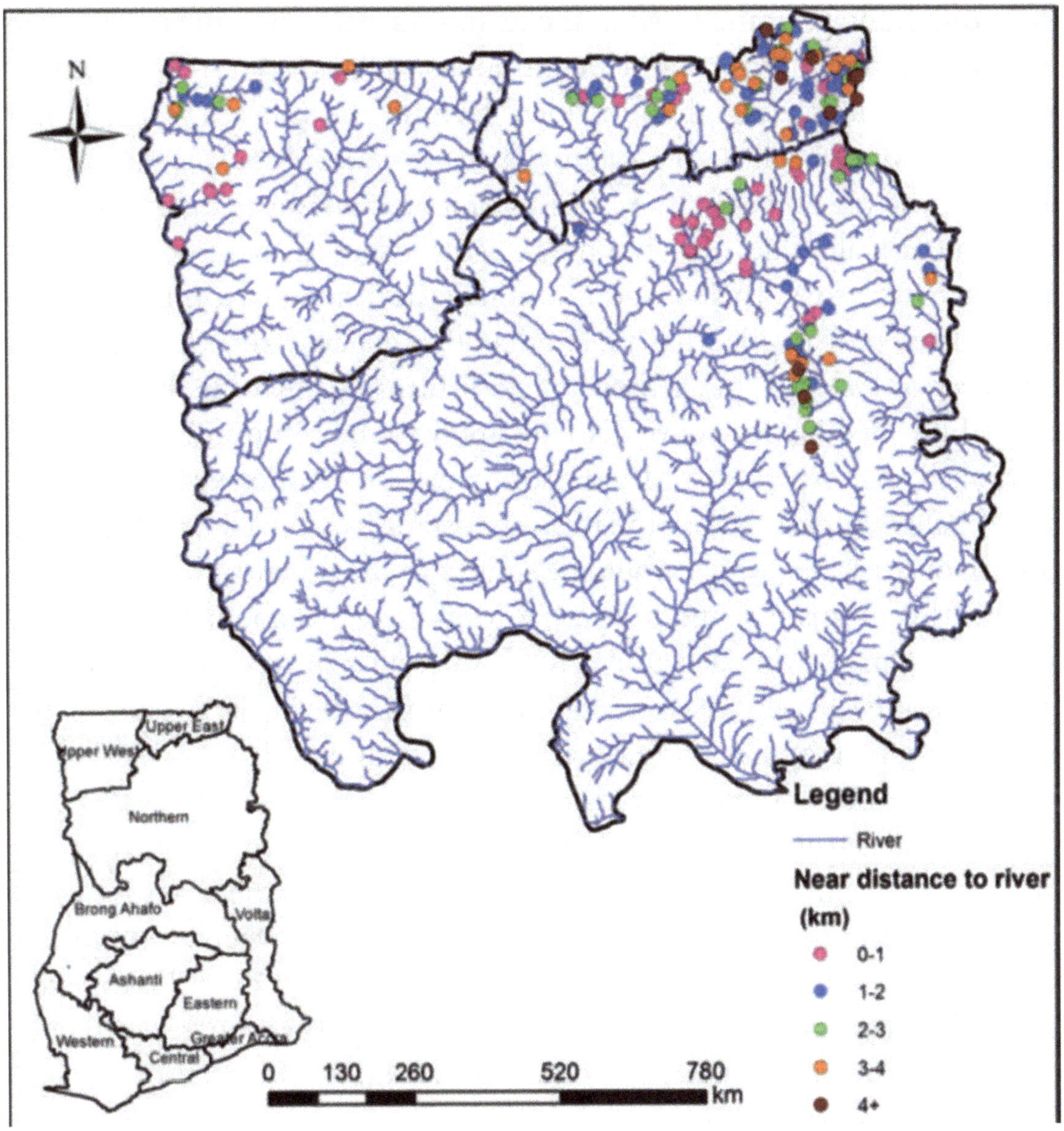

Fig. 9.1 Map of the study area. Source: Author

9.3.5 Statistical Methods

Deke (2014) in a Mathematica Policy Research brief suggested the Linear Probability Model (LPM) as appropriate compared to logistic model in calculating the impact of a binary outcomes in a randomized controlled trial setting.

The LPM has the ultimate advantage that the estimators can be directly interpreted as the marginal effect of covariates on the binary outcome. The main disadvantage of the LPM in the textbooks is that the true relationship between a binary outcome and a continuous explanatory variable is inherently nonlinear. That is, the functional form of the LPM is generally not correctly specified, which can lead to biased estimates of some parameters of interest. The reason is that the LPM assumes a constant marginal effect of covariate *X* for all values of *X*, but the marginal effect of *X* almost always varies with respect to *X*. This misspecification of the functional form often leads to predicted probabilities that are less than 0 or greater than 1 (Deke 2014). However, as Deke (2014) demonstrated, it turns out that the disadvantage of the LPM highlighted above does not apply to the context of randomized controlled trial experiment. The reason why LPM works well to estimate experimental impacts is that the treatment status is a binary variable, not a continuous variable, which would be subject to the potential bias described above. This means that the functional form concerns about LPM do not apply to estimating impacts under RCT, since all that is required is to estimate two prevalence rates; one for the treatment group and one for the control group (as opposed to estimating a different prevalence rate for every unique value of a continuous variable). A second reason that the LPM provides accurate estimates of experimental impacts is that any other covariates included in the impact model are uncorrelated with treatment status, which means that the impact estimate is unbiased regardless of whether the correct functional form is used to adjust for other (possibly continuous) covariates.

The LPM is simply the application of ordinary least squares to binary outcomes instead of continuous outcomes as follows:

$$\text{WTP}_i^{\text{OLS}} = \beta_0 + \beta_{\text{bid}}\text{Bid}_i + \beta_X X_i + \beta_T T_i + \beta_D D_i + \varepsilon_i$$

WTP_i is the binary response to the willingness to pay question for farmer i; Bid_i is the proposed price of the SI to farmer i; X_i is the vector of household characteristics; T_i is a vector of binary variables representing whether the respondent was issued drought index insurance last season or not; β_T and β_X are parameters representing mean marginal effects; D_i is the vector of district dummies and ε_i is the error term.

9.4 Results

9.4.1 Experimental Integrity: Balance Tests on Variables at Baseline

Table 9.5 presents the summary statistic and the balance test for the whole sample and for control and treatments groups with simple mean comparison *t*-tests at baseline and Table 9.6 presents the balance test for the WTP variable at baseline without and with compliance.

On average, the respondents are 46 years old and 83 percent of them do not have any formal education (Table 9.7). Male represents 52 percent of the sample. The household is composed of predominantly inactive members. The average household size is about 11 members with a dependency ration of 1.4. That is every active member of the household is in charge of more than one inactive members. On average, about six members of the household participate in agricultural labor. Households on average earn GHC 1334.00 from agriculture compared to the average total income of GHC 2889.00. Households typically obtain about 60 percent of their income from agriculture (compared to less than 6 percent on remittances). As shown in Table 9.8, nearly all (96 percent) are reliant on rainfall for crop production. Also, 97 percent of household own livestock. The average livestock endowment measured by tropical livestock units (TLU) is 3.43 (Table 9.5).

Landholdings in Ghana are typically small. Small farms predominate throughout the country, although they tend to be larger on average in the savannah zones, with land distribution more skewed closer to the coast. Average landholding in our sample is 6.20 acres (with more than 62 percent holding less than five acres) which is considerable higher than the national average of 5.6 acres (Chamberlin 2008). Household is reached by extension service officer about two times per season and the average walking time to the market is one hour.

On average, 53 percent of farmers experienced drought the previous cropping season and about 47 percent experienced at least three times and 91 percent experienced at least two times drought in the previous five cropping seasons. On average, 53 percent of farmers believe there will be

Table 9.5 Experimental integrity: balance tests on variables at baseline [3]

	Whole	Control	Treatment 1	Treatment 2	*t*-statistic	
Variables	(1)	(2)	(3)	(4)	(2) # (3)	(2) # (4)
Age	45.80	45.93	46	45.48	−0.07	0.45
	(12.73)	(13.15)	(12.01)	(13.04)		
Male	0.52	0.55	0.50	0.50	0.04	0.05
	(0.50)	(0.50)	(0.50)	(0.50)		
Household size	10.71	10.92	11.11	10.10	−0.19	0.81
	(6.65)	(5.73)	(8.05)	(5.86)		
Dependency ratio	1.39	1.24	1.25	1.68	−0.01	−0.44*
	(2.73)	(1.90)	(2.49)	(3.54)		
Total income (100)	23.89	25.03	22.63	24.02	2.41*	1.01
	(15.30)	(16.12)	(15.11)	(14.58)		
Agricultural income (100)	13.34	13.93	13.10	12.96	0.86	0.99
	(8.86)	(9.18)	(9.19)	(8.69)		
Remittance (100)	1.24	1.38	1.01	1.34	0.38**	0.04
	(2.23)	(2.35)	(2.17)	(2.34)		
Saving (100)	0.79	0.77	0.79	0.78	−0.02	−0.01
	(0.41)	(0.41)	(0.40)	(0.42)		
Loan received (100)	2.18	2.16	2.06	2.33	0.10	−0.17
	(2.72)	(2.89)	(2.64)	(2.64)		
TLU	3.43	3.52	3.30	3.48	0.22	0.39
	(3.47)	(3.57)	(3.38)	(3.45)		
Distance nearest market	1.05	1.00	1.03	1.11	−0.02	−0.10
	(0.77)	(0.70)	(0.74)	(0.87)		
Extension_ service	1.81	1.89	1.70	1.84	0.18	0.05
	(1.64)	(1.71)	(1.52)	(1.69)		
Drought 2014, Dummy	0.51	0.50	0.50	0.53	0.01	−0.05
	(0.50)	(0.50)	(0.50)	(0.50)		
Years of drought experiences	2.48	2.48	2.47	2.49	0.01	0.01
	(0.82)	(0.81)	(0.81)	(0.86)		
Drought_ perception	0.53	0.51	0.53	0.55	−0.02	−0.04
	(0.50)	(0.50)	(0.50)	(0.50)		
Help during drought	2.40	2.52	2.18	2.49	0.34*	0.03
	(2.39)	(2.49)	(2.14)	(2.50)		
Risk preferences						
Risk averse	0.39	0.36	0.44	0.36	−0.08*	0.003
	(0.49)	(0.48)	(0.50)	(0.48)		
Risk neutral	0.13	0.12	0.15	0.13	−0.03	−0.01
	(0.34)	(0.32)	(0.36)	(0.34)		
Risk loving	0.16	0.17	0.13	0.18	0.05	−0.01
	(0.37)	(0.38)	(0.33)	(0.39)		
Time preference	0.06	0.06	0.06	0.06	0.002	−0.003
	(0.06)	(0.05)	(0.45)	(0.07)		

(*continued*)

Table 9.5 (continued)

	Whole	Control	Treatment 1	Treatment 2	*t*-statistic	
Variables	(1)	(2)	(3)	(4)	(2) # (3)	(2) # (4)
Household labor	5.76	5.76	6.02	5.49	−0.26	0.27
	(4.01)	(3.84)	(4.87)	(3.39)		
Plots	3.95	4.01	3.82	3.91	0.28	0.19
	(2.22)	(2.40)	(2.10)	(2.15)		
Riceland	0.67	0.60	0.71	0.69	−0.11	−0.09
	(1.76)	(1,36)	(1.68)	(2.16)		
Landholding	6.20	6.09	6.51	5.99	−0.42	0.11
	(6.24)	(4.71)	(8.00)	(5.54)		
Observations	777	261	261	255		

Note: Standard deviation in parentheses.*Significant at 10 percent; **Significant at 5 percent; ***Significant at 1 percent

Table 9.6 Experimental integrity: WTP at baseline [4]

	Whole sample	Control	Treatment 1	Treatment 2	*t*-statistic	
Variables	(1)	(2)	(3)	(4)	(2) # (3)	(2) # (4)
WTP canal CSA	78.64	80.01	76.25	79.61	3.76	0.41
	(0.41)	(0.4)	(0.4)	(0.4)		
WTP drip CSA	80.69	84.29	78.16	79.61	6.13	4.68
	(0.39)	(0.36)	(0.41)	(0.4)		
Observations	777	261	261	255		
With compliance						
WTP canal CSA	77.4	77.95	73.53	80.55	4.42	−2.6
	(0.42)	(0.42)	(0.44)	(0.40)		
WTP drip CSA	79.36	80.31	77.21	80.55	3.1	−0.24
	(0.4)	(0.4)	(0.42)	(0.4)		
Observations	407	127	136	144		

Note: Standard deviation in parentheses. *Significant at 10 percent; **Significant at 5 percent; ***Significant at 1 percent

Table 9.7 Household level of education [5]

Level of education	Freq.	Percent
No education	648	83.40
Primary school	40	5.15
Middle school	46	5.92
High school/university	43	5.53
Total	777	100.00

Table 9.8 Access to irrigation [6]

Access to irrigation	Freq.	Percent
No	743	95.62
Yes	34	4.38
Total	777	100.00

drought next coming cropping season. Farmers could call upon two to three members for help if there is drought.

9.4.2 Impact of Drought Index Insurance on the Demand for Supplemental Irrigation

We report the results of the effect of drought index insurance on SI in three steps: first, the relationship between demand of SI and treatments is presented via simple mean comparison. Second, we report the result of the LPM. Third, we analyze whether the impact differs when allowing for respondents heterogeneity.

Farmers who actually received the treatment represent 79.22 percent of farmers initially assigned to the treatment. Treatment was contingent on the receiving of loans. Therefore, farmers initially assigned to control and treatments who could not get credit from the banks were taken out of the analysis. This does not have any major implication in the power of our analysis as we considered 50 percent of compliance in the calculation of the sample size. Besides, as presented in the Table 9.9, compliance is almost equally distributed across groups.

To test whether demand for SI was higher or lower among treatments, we first looked at simple mean outcomes post-intervention. The randomization and the fact that the control and treatment samples are well balanced in the observed characteristics imply that a simple comparison of mean outcomes post-intervention will likely provide an unbiased estimate of intervention impacts. However, we also control for other observed socioeconomic characteristics in order to reduce idiosyncratic variation and to improve the power of the estimates.

Table 9.10 presents the results of a simple comparison of mean outcomes post-intervention. We observe that the two treatment groups have higher demand for SI compared to the control group. In treatment 1 (drought index insurance with farmer as policy holder), the demand of SI increased by 7.8 percent for canal irrigation and by 12.8 percent for drip irrigation. In treatment 2 (drought index insurance with bank as policy holder), the demand for SI increased by 13.5 percent for canal irrigation and by 18.4 percent for drip irrigation. The differences between control and treatments groups are all significant. Although treatment 2 appears to have greater effect than treatment 1 in both canal and drip irrigation, their differences are not significant. Also, we could not find any significant difference of impact between canal and drip irrigation, except in the control group where the difference is significant at 5 percent level.

For the impact with covariates, we estimated a LPM of WTP for SI with the treatment variable being a dummy indicating whether the individual was in one of the treatment groups or in the control group. The covariates are socioeconomic characteristics measured during the follow-up survey which took place just after the intervention. The specific variables included in the model were those highly correlated with the dependent variables (dummy WTP) in the control group (Gertler 2004).

We first considered only the WTP bid as the only covariate. Table 9.11 provides the result of the treatments effect using LPM specification.

Finally, we added the other covariates as presented in Table 9.12. We find similar magnitude effects of drought index insurance on the WTP for SI as those found in Table 9.11.

Table 9.9 Actual (preliminary) farmers into treatment and control groups by regions [7]

Treatment	Northern	Upper West	Upper East	Total
Control	57 (103)	12 (27)	69 (131)	138 (261)
Treatment 1	75 (96)	30(33)	78 (132)	183 (261)
Treatment 2	66(99)	15(24)	90 (132)	171 (255)
Total	198 (298)	57 (84)	237 (395)	407 (777)

Table 9.10 Mean treatment effect [8]

	Control	Treatment 1	Treatment 2	Impact (*t*-statistic)		
	(1)	(2)	(3)	(1) # (2)	(1) # (3)	(2) # (3)
Canal SI	72.5	80.3	86.0	−7.8*	−13.5***	−5.7
Drip SI	68.1	80.9	86.5	−12.8***	−18.4***	−5.6
Diff (*t*-statistic)	4.4**	−0.6	−0.5	5	4.9	−0.1

The observations are percentage. $^{***}p < 0.01$, $^{**}p < 0.05$, $^{*}p < 0.1$

Table 9.11 Treatment effect with bid WTP LPM [9]

	Canal SI			Drip SI		
	(1)	(2)	(3)	(4)	(5)	(6)
WTP_bid	−0.058***	−0.054***	−0.043***	−0.057***	−0.056***	−0.037***
	(0.007)	(0.007)	(0.007)	(0.007)	(0.007)	(0.007)
Insurance_farmer	0.069*			0.118***		
	(0.043)			(0.044)		
Insurance_bank		0.121***			0.170***	
		(0.042)			(0.043)	
Insurance_farmer_bank			0.052			0.053
			(0.037)			(0.037)
Constant	1.867***	1.787***	1.646***	1.812***	1.792***	1.526***
	(0.135)	(0.141)	(0.127)	(0.133)	(0.142)	(0.123)
Observations	321	309	354	321	309	354
R-squared	0.205	0.196	0.133	0.205	0.223	0.101
Log likelihood	−141.1	−122	−130	−148.2	−125.6	−131.6

The observations are percentage. ***$p < 0.01$, **$p < 0.05$, *$p < 0.1$

These results cohere with the complementarity effect hypothesis of the effect of drought index insurance on SI. The complementarity nature of drought index insurance and SI that we found can be comprehended in two ways. First, farmers with drought index insurance might think of implicitly insuring the cost of irrigation. Drought index insurance have been widely studied in the past two decades in developing countries, yet its take up is still very poor with demand disappearing as soon as the subsidy is removed or the pilot project is terminated. Second, farmers probably do not find drought index insurance as stand-alone drought risk management instrument interesting enough especially that drought index insurance does not cover the actual loss and also the famous basis risk problem. The indemnity also appears very small to allow farmers to smooth their consumption during drought time especially that the trigger of the insurance is likely to be associated with the rise of the price of stable food in the local market. Farmers usually have high incentive to protect their yield compared to any other objective. This is because farmers heavily rely on their own production for consumption and usually express very strong resistance to buy any stable foods that they think they should have had from their farms. Therefore, insurance in this case only provides farmers with the opportunity to afford SI which is the drought risk management tools that they really prefer because it helps increase their production. Also, there is a strong correlation between the cost of irrigation or the disruption of irrigation and the likelihood of insurance being triggered. As the severity of drought increases, the cost of irrigation, the disruption of allocation of water for SI and the probability of insurance payout increase as well.

Mafoua and Turvey (2003) found similar result. The authors employed an economic model to analyze the tradeoff between the loss in revenues from unirrigated crops during drought and the cost of irrigation to preserve yields in periods of drought and simulate drought index insurance in both scenarios. They then came to the conclusion that rainfall insurance can be used to insure against the cost of irrigation.

Table 9.12 Treatment effect with covariate LPM

	Canal SI			Drip SI		
	(1)	(2)	(3)	(4)	(5)	(6)
WTP_bid	−0.059***	−0.054***	−0.044***	−0.058***	−0.056***	−0.037***
	(0.007)	(0.007)	(0.006)	(0.007)	(0.007)	(0.006)
Age	−0.000	0.001	0.001	−0.001	0.001	0.002
	(0.002)	(0.002)	(0.002)	(0.002)	(0.002)	(0.002)
Education	0.038*	0.044**	0.023	0.061***	0.042*	0.024
	(0.021)	(0.020)	(0.015)	(0.020)	(0.023)	(0.017)
Agricultural income	0.008***	0.006**	0.007***	0.006***	0.004	0.006***
	(0.002)	(0.003)	(0.002)	(0.002)	(0.003)	(0.002)
Loan received	0.007	−0.001	0.023***	0.005	−0.002	0.018***
	(0.007)	(0.007)	(0.006)	(0.007)	(0.007)	(0.006)
Remittance	−0.011	0.004	−0.004	−0.012	0.005	0.003
	(0.010)	(0.009)	(0.008)	(0.010)	(0.009)	(0.007)
Drought last season	0.127***	0.021	0.070*	0.130***	0.015	0.054
	(0.041)	(0.042)	(0.037)	(0.042)	(0.043)	(0.037)
Nearer river	−0.043	−0.049	−0.030	0.021	0.006	0.021
	(0.045)	(0.049)	(0.045)	(0.047)	(0.049)	(0.044)
Extension services	0.036**	0.013	0.001	0.025*	0.012	−0.002
	(0.014)	(0.012)	(0.011)	(0.014)	(0.013)	(0.011)
Landowned	0.005**	0.000	0.004**	0.004	0.000	0.003
	(0.002)	(0.005)	(0.002)	(0.002)	(0.005)	(0.003)
Share rice land	−0.020**	−0.001	−0.006	−0.012	0.004	−0.001
	(0.009)	(0.011)	(0.007)	(0.010)	(0.011)	(0.007)
Insurance_farmer	0.088**			0.136***		
	(0.042)			(0.043)		
Insurance_bank		0.119***			0.167***	
		(0.042)			(0.043)	

Insurance_farmer_bank			0.026			0.025
			(0.036)			(0.037)
Canal asked 1st	0.005	0.013	0.018	–	–	–
	(0.041)	(0.041)	(0.036)			
Drip asked 1st				0.044	0.019	0.015
				(0.043)	(0.042)	(0.037)
Constant	1.628***	1.635***	1.402***	1.586***	1.627***	1.253***
	(0.157)	(0.164)	(0.149)	(0.160)	(0.167)	(0.141)
Observations	321	309	354	321	309	354
R-squared	0.309	0.232	0.224	0.292	0.249	0.173
Log likelihood	−118.6	−115	−110.5	−129.8	−120.2	−116.9

The observations are percentage. ***$p < 0.01$, **$p < 0.05$, *$p < 0.1$

9.5 Conclusions

Index insurance and SI have the potential to hedge drought risk and as such they offer risk-averse farmers the opportunity to invest more on their farms. This study assess via a three-year randomized controlled trial experiment the effect of index insurance on the demand for SI among smallholder farmers. The study found the demand for SI canal (drip) among farmers who received insurance as policy holders to be 7.8 (12.8) percent greater than in the control group, and the demand for SI canal (drip) among farmers who received insurance with their banks as policy holders to be 13.5 (18.4) percent greater than in the control group.

It is, therefore, worthy investing in SI technology because it improves the overall profile of the profit distributions of farmers. But the only inconvenience is the risk of no allocation of water and the high cost of alternative solutions involved during severe drought years. Hence, coupled SI with drought index insurance for long dry spell or severe drought might help reduce the high cost of irrigation during severe drought and reduce the cost of insurance and basis risk that undermine the take up of insurance in developing countries.

Notes

1. Throughout this chapter, we focus our attention on a specify case of weather index insurance called rainfall or drought index insurance.
2. Many other outcome variables were investigated including: agricultural loan defaults; access to credit; interest rates charged and uptake of improved production technologies.
3. Annual potential evapotranspiration is about 2000 mm in the north.

References

Alhassan, M., Loomis, J., Frasier, M., Davies, S., & Andales, A. (2013). Estimating Farmers' Willingness to Pay for Improved Irrigation: An Economic Study of the Bontanga Irrigation Scheme in Northern Ghana. *Journal of Agricultural Science, 5*(4), 31–43.

Arrow, K., Solow, R., Portney, P. R., Leamer, E. E., Radner, R., & Schuman, H. (1993). Report of the NOAA Panel on Contingent Valuation. *Federal Register, 58*(10), 4601–4614.

Barham, E. H. B., Robinson, J. R. C., Richardson, J. W., & Rister, M. E. (2011). Mitigating Cotton Revenue Risk Through Irrigation, Insurance, and Hedging. *Journal of Agricultural and Applied Economics, 43*(4), 529–540.

Bloom, H. S. (1995). Minimum Detectable Effects: A Simple Way to Report the Statistical Power of Experimental Designs. *Evaluation Review, 19*(5), 547–556.

Bloom, H. S. (2006, August). The Core Analytics of Randomized Experiments for Social Research. *MDRC Working Papers on Research Methodology* (1–41). New York: MDRC.

Buchholz, M., & Musshoff, O. (2014). The Role of Weather Derivatives and Portfolio Effects in Agricultural Water Management. *Agricultural Water Management, 146*(February), 34–44.

Chamberlin, J. (2008). It's a Small World After All: Defining Smallholder Agriculture in Ghana. *IFPRI Discussion Papers*.

Chandrasekaran, K., Devarajulu, S., & Kuppannan, P. (2009). Farmers' Willingness to Pay for Irrigation Water: A Case of Tank Irrigation Systems in South India. *Water, 1*(1), 5–18.

Chantarat, S., Mude, A. G., & Barrett, C. B. (2009). *Willingness to Pay for Index Based Livestock Insurance: Results from a Field Experiment in Northern Kenya*. Ithaca, NY: Cornell University.

Dalton, T. J., Porter, G. a., & Winslow, N. G. (2004). Risk Management Strategies in Humid Production Regions: A Comparison of Supplemental Irrigation and Crop Insurance. *Agricultural and Resource Economics Review, 33*(2), 220–232.

Deke, J. (2014). *Using the Linear Probability Model to Estimate Impacts on Binary Outcomes in Randomized Controlled Trials*. Mathematica Policy Research, HHS Office of Adolescent Health.

Dietz, A. J., Ruben, R., & Verhagen, A. (2004). *The Impact of Climate Change on Dryland With a Focus on West Africa, 488*. Retrieved from http://www.ebooks.kluweronline.com.

Duflo, E., Glennerster, R., & Kremer, M. (2008). Using Randomization in Development Economics Research: A Toolkit. *Handbook of Development Economics, 4*, 3895–3962.

Elagib, N. A. (2014). Development and Application of a Drought Risk Index for Food Crop Yield in Eastern Sahel. *Ecological Indicators, 43*, 114–125.

EPA. (2007). *Climate Change and the Ghanaian Economy*. Environmental Protection Agency (EPA), Policy Advise Series.

EPA. (2012). *Climate Change Impact: Why must Ghana Worry?* Environmental Protection Agency (EPA), Policy Advise Series.

FAO. (2007). Climate Change, Water and Food Security. FAO, Rome. Retrieved from ftp://ftp.fao.org/docrep/fao/010/i0142e/i0142e07.pdf.

Farrin, K., & Miranda, M. J. (2015). A Heterogeneous Agent Model of Credit-Linked Index Insurance and Farm Technology Adoption. *Journal of Development Economics, 116*, 199–211.

Foudi, S., & Erdlenbruch, K. (2012). The Role of Irrigation in Farmers' Risk Management Strategies in France. *European Review of Agricultural Economics, 39*(3), 439–457.

Fox, P., & Rockström, J. (2000). Water-Harvesting for Supplementary Irrigation of Cereal Crops to Overcome Intra-Seasonal Dry-Spells in the Sahel. *Physics and Chemistry of the Earth, Part B: Hydrology, Oceans and Atmosphere, 25*(3), 289–296.

Gertler, P. (2004). Do Conditional Cash Transfers Improve Child Health? Evidence from PROGRESA's Control Randomized Experiment. *The American Economic Review, 94*(2), 336–341.

Hellmuth, M. E., Moorhead, A., Thomson, M. C., & Williams, J. (2007). *Climate Risk Management in Africa: Learning from Practice.* International Research Institute for Climate and Society (IRI), Columbia University, New York.

Hesselberg, J., & Yaro, J. A. (2006). An Assessment of the Extent and Causes of Food Insecurity in Northern Ghana Using a Livelihood Vulnerability Framework. *GeoJournal, 67*(1), 41–55.

Laube, W., Awo, M., & Schraven, B. (2008). *Erratic Rains and Erratic Markets: Environmental Change, Economic Globalisation and the Expansion of Shallow Groundwater Irrigation in West Africa.*

Lin, S., Mullen, J. D., & Hoogenboom, G. (2008). Farm-Level Risk Management Using Irrigation and Weather Derivatives. *Journal of Agricultural and Applied Economics, 40*(2), 485–492.

Mafoua, E., & Turvey, C. (2003). *Weather Insurance to Protect Specialty Crops Against Costs of Irrigation in Drought Years.* Paper Prepared for Presentation at the Annual Meeting of the American Agricultural Economics Association, July 27–30, Montreal.

Miranda, M. J. (1991). Area-Yield Crop Insurance Reconsidered. *American Journal of Agricultural Economics, 73*(2), 233–242.

MoFA. (2007). *Food and Agriculture Sector Development Policy.* FASDEP II-Ministry of Food and Agriculture.

Oweis, T., & Hachum, A. (2006). Water Harvesting and Supplemental Irrigation for Improved Water Productivity of Dry Farming Systems in West Asia and North Africa. *Agricultural Water Management, 80.*

Oweis, T., & Hachum, A. (2012). *Supplemental Irrigation a Highly Efficient Water – Use Practice.* Syria: International Centre for Agricultural Research in the Dry Areas.

Rockström, J., Karlberg, L., Wani, S. P., Barron, J., Hatibu, N., Oweis, T., et al. (2010). Managing Water in Rainfed Agriculture – The Need for a Paradigm Shift. *Agricultural Water Management, 97*(4), 543–550.

Shiferaw, B., Tesfaye, K., Kassie, M., Abate, T., Prasanna, B. M., & Menkir, A. (2014). Managing Vulnerability to Drought and Enhancing Livelihood Resilience in sub-Saharan Africa: Technological, Institutional and Policy Options. *Weather and Climate Extremes, 3,* 67–79.

van de Giesen, N., Liebe, J., & Jung, G. (2010). Adapting to Climate Change in the Volta Basin, West Africa. *Current Science, 98*(8), 1033–1037.

Yaro, J. A. (2013). The Perception of and Adaptation to Climate Variability/Change in Ghana by Small-Scale and Commercial Farmers. *Regional Environmental Change, 13*(6), 1259–1272.

Zongo, B., Diarra, A., Barbier, B., Zorom, M., Yacouba, H., & Dogot, T. (2015). Farmers' Perception and Willingness to Pay for Climate Information in Burkina Faso. *International Journal of Food and Agricultural Economics, 3*(1), 101–117.

Part III

Promoting Agro-Industrialization

10

Does the Development of the Agricultural Sector Affect the Manufacturing Sector?

Namalguebzanga Christian Kafando

10.1 Introduction

In the mid-twentieth century, most African countries proclaimed their independence and resolved to progress economically from low value-added primary activities to high value-added manufacturing activities. This process which is referred to as "industrialization" was considered as the main means of evolving from developing to developed or industrialized countries. In 1986, consensus on this process also emerged within the Organization of African Unity (OAU) in Africa's Priority Programme for Economic Recovery 1986–1990. Soon after independence, African countries tried to build an industrial sector-driven economy. Unfortunately, this strategy failed especially due to very strong State interference in the economic sector and a difficult global context (Hughes 1984; Hawkins 1986).

N. C. Kafando (✉)
United Nations Industrial Development Organization, Abuja, Nigeria

A. Shimeles et al. (eds.), *Building a Resilient and Sustainable Agriculture in Sub-Saharan Africa*, https://doi.org/10.1007/978-3-319-76222-7_10

This explains why Africa is the only continent that has not benefited from an expansion of the industrial development model but also the most marginalized continent with regard to manufacturing output and world trade (UNCTAD 2011). Very recently, however, the United Nations Economic Commission for Africa and the African Union seemed to have underscored the need to revamp industrial policy in Africa. "Given that most African countries now have a comparative advantage in commodities, industrialization based on the exploitation of natural resources is a means for developing regional value chains on the continent, and African countries should take advantage of this opportunity" (ECA and AUC 2013; UNCTAD 2013).

This study is carried out within this framework for reasons presented earlier and focuses on the contribution of the agricultural sector to industrial development. This sector has already been the subject of a consensus at the continental level through the 2003 Maputo (Mozambique) Declaration. The objective of this Agreement, which was materialized by the adoption of the Comprehensive Africa Agriculture Development Programme (CAADP), is to improve the contribution of the agricultural sector to the economic growth of African countries. This objective is especially commendable as this sector is vital to industrial development.

The main objective of this study is to understand the contribution of agricultural activities to the development of the manufacturing sector in African countries but also to determine if and how all African countries can take advantage of the industrial development process. The study is divided into four parts. Part I presents a review of the relevant literature and identifies the variables used for the study. Part II presents the analytical framework and a statistical description of the links between the agricultural and manufacturing sectors in Africa, while making a distinction between the different regions. This distinction is necessary because the industrial sector comprises several sectors, namely manufacturing, mining and oil (Lewis 1954). The study focuses on the manufacturing industry because it is the only sector that can be affected by the agricultural sector. Part III analyses the contribution of agricultural activities to the development of the manufacturing sector in African economies using econometrics. In other words, the study examines the relationship between the levels of agriculture value added and those of the manufacturing sector in African countries, taking into account growth factors (physical capital, human capital and economic openness) and the quality of governance. The regressions made is based on fixed effects models, double fixed effects models and the generalized least-squares estimator. Part IV presents and interprets the implications in order to draw conclusions in terms of economic policies.

10.2 Literature Review

Concerning the development outlook of African countries, we have seen that some studies suggest that these countries should only specialize in the production of natural resources in which they have a comparative advantage (Wood and Mayer 2001; Park and Lee 2006; Mayer and Fajarnes 2008). On the other hand, other authors have upheld that industrialization is necessary because it constitutes the key element in efforts to close the economic gap (Cornwall 1977; Tregenna 2007; Szirmai and Verspagen 2011). Agriculture alone cannot steer countries on the path of strong growth because, according to classical economic theories, the value added of output per capita generated by the agricultural sector is less than the one generated by the industrial sectors although it is vital for industrial development (Lewis 1954).

The level of development of the agricultural sector can be considered as a mean of developing the industrial sector. This feature also has many advantages. "One of the most striking features of developing countries is a weak linkage between agriculture and industry. That is why agriculture cannot experiment new inputs such as fertilizers and machinery" (Abdelmalki and Mundler 1995). After examining the importance of agriculture in the development of manufacturing activities in the economy on a global sample and taking into account some regions using dummy variables, Shifa (2011) observed that this impact is positive and proportional to the share of agriculture in the economy. Mellor (1966) argues that an increase in the size of the agricultural sector promotes natural transformation. This enables the economy to evolve from a situation dominated by a slow-growing agricultural sector to another dominated by a fast-growing non-agricultural sector.

Many authors have pointed out the importance of the agricultural sector for the take-off of industrial development (Bairoch 1971; Kuznets 1966; Fei and Ranis 1969). Mellor (1966) explains that change in economic structure is due to the size of the agricultural sector rather than its growth rate which is slow. According to him, the improvement of human capital and dissemination of knowledge promotes agricultural development through the accumulation and distribution of capital among sectors, the specialization of economies and economic openness. Balassa (1979) argues that differences in physical and human capital lead to differences in the performance of the manufacturing sectors of economies. According to him, population growth, which is often regarded as a drag on development, veils a combination of human creativity and the accumulation of scientific knowledge which help to improve technology. The use of technology in the production system leads to an increase in agricultural sector productivity. This productivity is subsequently enhanced by the advantages related to specialization in a given sector of production. There are two advantages of specialization: it reduces transaction costs and increases the volume of trade.

In reality, most of the explanations given for the impact of the size of the agricultural sector on the industrial sector are based on the concept of surplus. Some authors (Lewis 1954; Fei and Ranis 1969; Abdelmalki and Mundler 1995) have also tried to explain this change using the concept of agricultural surplus which is transferred to the manufacturing sector. Agricultural surplus may be in the form of workforce, additional production or additional income.

The concept of agricultural surplus (Lewis 1954) may vary depending on whether we are addressing supply and demand, the labour market, the self-financing of the economy or global trade issues (Abdelmalki and Mundler 1995). Thus, four definitions of surplus can be proposed.

Concerning the supply of and demand for agricultural products, an increase in production helps to meet the needs of the farm population and the rest of the population. Part of the agricultural surplus generated can also be used or processed by the industrial sector. An increase in the supply of agricultural products leads to a drop in the prices of those products which also bring about a reduction in wage costs in all sectors of the economy.

Regarding the labour supply and demand market, an increase in agricultural sector productivity will release surplus labour which will be available to other economic sectors.

Breakthroughs in agriculture help to mobilize “forced” or “voluntary” savings that could lead to economic self-financing (Abdelmalki and Mundler 1995). Savings are considered as forced when they are generated in the form of taxes or levies on proceeds of the sale of agricultural products. The State plays a key role by forcibly mobilizing financial resources which are reinjected into the economy. Savings are voluntary when they are made by the farmers themselves in the sense that surplus profit is reinvested through the acquisition of means to improve production or productivity or simply as a means for future consumption. According to Abdelmalki and Mundler (1995), in both cases, farmers' actions help to stimulate demand.

Through exports of agricultural products, local economic actors earn foreign exchange which is used to finance the industrial sector (Mellor 1966; Abdelmalki and Mundler 1995). In this case, export earnings would depend on terms of trade corresponding to purchasing power (Abdelmalki and Mundler 1995). The terms of trade are favourable when export prices increase faster than import prices and unfavourable otherwise.

Besides these elements, the way a country is governed can determine its capacity to mobilize resources. Auty (2000) argues that governance that allows for the implementation of reforms by encouraging the development of manufacturing sector activities enables the economy to experience sustainable and equitable growth. In contrast, a government controlled by a group of people does not allow the establishment of such a virtuous circle. According to this model, the quality and type of governance determine the capacity of an economy to industrialize. Collier (2002) believes that poor governance is linked to dependence on primary commodities. Though this dependence can impede the development of manufacturing industries, bad institutions can lead to poor performance.

This literature review has helped to identify the size of the agricultural sector as a vehicle of industrial development and the factors (investment, human capital, governance quality and economic openness) that contribute to industrial development.

10.3 Analytical Framework

To measure the size that incorporates the effects of productivity and agricultural sector surplus, we considered the per capita value added created by this sector (agrvapc, WB). To take into account the investment made in the economy as a whole, we considered gross fixed capital formation per capita (gfcfpc, UNCTAD). Since human capital can boost manufacturing and agricultural sector productivity, we use an index to measure the development potential of human capital that is the "Human Asset Index (HAI)" (UN DESA, FERDI, 2011). Since proper resource allocation depends on governance and can promote or hinder manufacturing sector development, we conceptualized it by measuring good governance, denoted as "qogov" (ICRG). Economic openness is considered through the Penn World Table 7 "openc" variable. Lastly, as dependent variable, instead of using the industrial sector as a whole, we opted for the manufacturing sector which lies at the very heart of the ripple effects linked to the industrial sector. This level of manufacturing is just like that of the agricultural sector which is measured by its value added per capita (manufacturing value added per capita—MVAPC, UNCTAD). The period covered by the data runs from 1980 to 2009, which means 30 years.

Thus, in light of the various contributions of the literature presented earlier, we retain the size of the agricultural and manufacturing sectors rather than their growth rate.

We present two types of models. Unlike the first model, the second includes temporal dummies (f_t) to consider temporary shocks.

The models can be presented as follows with X representing the vector of control variables:

$$\text{mvapc}_{i,t} = \text{agrvapc}^{\beta}_{i,t-1} \times X^{\delta}_{i,t} \tag{10.1}$$

When expressed in logarithmic form, Eq. 10.1 is as follows:

$$\ln \text{mvapc}_{i,t} = \alpha + \beta \ \text{agrcapc}_{i,t-1} + \delta X_{i,t-1} + f_t + \varepsilon_{i;t} \tag{10.2}$$

where MVAPC represents the value added per capita, *a* the constant of the model, β, δ the coefficients of the variables measuring the effects of agricultural value added and control variables, namely the level of investment per capita, a variable estimating the value of human capital, economic openness, governance quality and the fixed effects of the period on the dependent variable. The notation $t - 1$ means that we have considered lagged variables as current variables could be endogenous to the dependent variable. Agriculture does not have an immediate effect on manufacturing activities. The effects of an increase in the size or scale of agricultural production take a long time to manifest themselves (Studennund 2000). "Any change in an agricultural market, such as an increase in the price that the farmer can earn for providing cotton, has a lagged effect on the supply of that product" (Studennund 2000). Then, where there is change in demand in the agricultural products processing sector during the period t, agricultural sector producers will adapt their supply during the period $t + 1$. Thus, the use of the lagged value of agricultural value added per capita helps to take these aspects into account. Thus, it is easy to imagine, for example, that agricultural production during $t - 1$ can have an impact on the production of manufactured goods related to the processing of agricultural products. The same applies to levels of investment, education and exports during $t - 1$. These reasons help to justify the use of lagged variables as a means to address endogenous phenomena like Barro (1998) did in his studies on growth. Thus, all independent variables will be modified by their lagged value in a given year, except our measurement of governance quality.

We estimate the model below using a sample of 37 African countries, making sure to distinguish the regions (excluding countries for which we do not have sufficient observations[1] on governance quality) (Table 10.1 shows descriptive statistics). For all the samples studied, the data are divided into three-year periods, which represent a total of ten periods. The notation "*l*" means that the variables are lagged.

$$\ln \text{mvapc}_{i,t} = \alpha + \beta \ l.\ln \text{agrcapc}_{i,t-1} + \delta X_{i,t-1} + f_t + \varepsilon_{i;t} \quad (10.3)$$

Table 10.1 Country classification by region

	Indicators	Manufacturing value added per capita in USD		Agricultural value added per capita in USD	
Regions	Countries	Median values	Regional median	Median values	Regional median
Southern Africa	South Africa	808	233	122	121
	Malawi	20		65	
	Namibia	351		316	
	Botswana	175		128	
	Zambia	61		134	
	Zimbabwe	52		71	
	Mozambique	21		66	
North Africa	Algeria	169	212	206	217
	Egypt	154		163	
	Morocco	256		262	
	Libya	374		184	
	Tunisia	400		278	
Central Africa	Angola	90	28	206	**206**[a]
	Cameroon	166		178	
	Central African Republic	28		177	
	Gabon	19		251	
	Equatorial Guinea	8		252	
East Africa	Kenya	58	29.5	131	148
	Somalia	6		165	
	Sudan	37		206	
	Tanzania	22		96	
West Africa	Côte d'Ivoire	155	43	283	164
	Benin	39		161	
	Ghana	78		205	
	Guinea-Bissau	46		210	
	Liberia	14		167	
	Senegal	98		117	
	Sierra Leone	20		159	
	Togo	32		152	

Sources: Calculations made by the author based on data derived from the United Nations Conference on Trade and Development (UNCTAD) and Penn World Table (PWT) 7.0; Alan Heston, Robert Summers and Bettina Aten, Penn World Vision 7.0, Center for International Comparisons of Production, Income and Prices at the University of Pennsylvania, May 2011

[a]206 is the median value of the Agricultural value added per capita in USD in Central Africa

Table 10.2 Generalized Hausman test

	4	5	6
Africa	320.42*** (0.0000)	380.90*** (0.0000)	411.89*** (0.0000)
Southern Africa	77.88*** (0.0000)	63.17*** (0.0000)	59.64*** (0.0000)
North Africa	56.25*** (0.0000)	82.22*** (0.0000)	114.76*** (0.0000)
East Africa	53.55*** (0.0000)	353.25*** (0.0000)	447.58*** (0.0000)
Central Africa	80.19*** (0.0000)	223.50*** (0.0000)	232.53*** (0.0000)
West Africa	149.83*** (0.0000)	158.00*** (0.0000)	162.93*** (0.0000)

Robust *p* value in parentheses
*** $p<0.01$, ** $p<0.05$, * $p<0.1$

We identified five subsamples in all, with each subsample representing a region. Thus, we make a distinction between North Africa, Southern Africa, West Africa, Central Africa and East Africa.

As the right model to use on our different panels could not be determined by the usual Hausman test, we used a generalized Hausman test which is in fact an increased regression that is asymptotically equivalent to the Hausman test (Wiggins 2003). The results of the test (Table 10.2) suggest that the fixed effects model is preferable to the random effects model for all samples. Later on, we used double effects fixed models, which are models that take into account countries characteristics and the effects that are specific to each period. Since the use of lagged values for some variables does not help to eliminate all the probability of endogenous factors in so far as hysteresis effects may still exist, we used the dynamic panel estimator of the generalized moments method (GMM) in first two-step differences based on the works of Arellano and Bond (1991), Arellano and Bover (1995) and Blundell and Bond (1998) (Roodman 2006). This is an estimator that is especially adapted to samples with many individuals (37 countries in our sample) and few periods (ten periods of three years each in our study). This estimator helps to eliminate order 1 autocorrelation, making it possible to validate the exogenous factors of the instruments used. This method was therefore used to estimate the following dynamic model (Eq. 10.4):

$$\begin{aligned}\ln \text{mvapc}_{i,t} &= \alpha + \alpha \ l.\ln \text{agrcapc}_{i,t-1} + \beta \ l.\ln \text{agrcapc}_{i,t-1} \\ &+ \delta X_{i,t-1} + f_t + \varepsilon_{i;t}\end{aligned} \tag{10.4}$$

We applied the Windmeijer correction method (2005) to minimize residuals (Roodman 2006). To verify if error terms do not correlate, we also carried out a Sargan/Hansen test. Unfortunately, given the size of the sample at the regional level, it was not possible for us to carry out this robustness test for the different regions. However, we were able to do so for the bulk sample (i.e., the 37 countries).

Before interpreting the results of the models described earlier, we present the statistical analysis of the data used in this empirical study.

10.4 Description of Variables

To better understand the relationships between our two variables of interest in Africa, it is necessary to compare manufacturing value added per capita with agriculture value added per capita. The data concerning value added by sector and level of investment are expressed in 2005 dollar per capita.

The curves in Graph 10.1 suggest that manufacturing value added is lower than agricultural value added. This shows that the industrial sector is not the key economic sector in Africa.

It should, however, be noted that during the 1998–2009 period, the per capita value added of these two sectors seems to be correlated. This can also be interpreted as a change in trend in recent years. One of the reasons for such change may be the use of agricultural surplus in the processing sector.

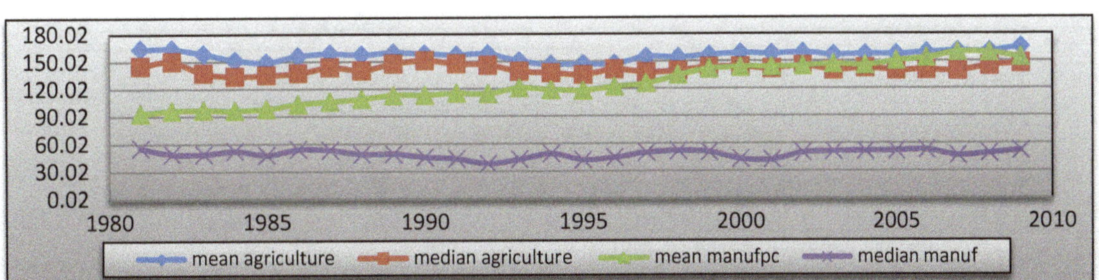

Graph 10.1 Manufacturing and agricultural value added per capita. Sources: Calculations made by the author based on data derived from the United Nations Conference on Trade and Development (UNCTAD) and Penn World Table (PWT) 7.0; Alan Heston, Robert Summers and Bettina Aten, Penn World Vision 7.0, Center for International Comparisons of Production, Income and Prices at the University of Pennsylvania, May 2011

However, the explanation is much more complex than it seems. Concerning the manufacturing sector, this same graph suggests that its value added per capita increased on average during the entire period. This difference between the median and the sector average very clearly suggests that the countries in the sample did not experience similar trends.

These initial differences do not bar us from asserting that there is a positive correlation between the two sectors (Graph 10.2). We opted to further analyse the data to better understand the differences that may exist at the regional level (Graph 10.3).

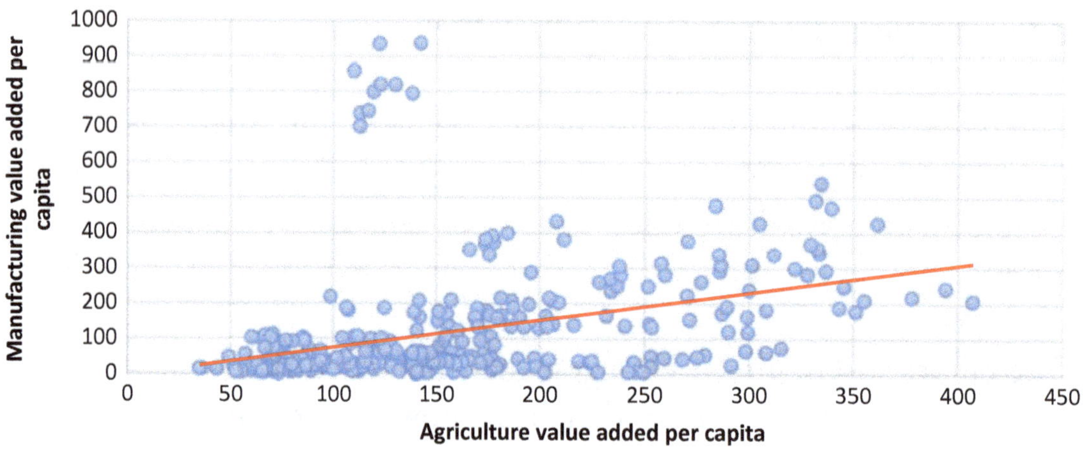

Graph 10.2 Linear trend in agricultural and manufacturing sector value added. Sources: Calculations made by the author based on data derived from the United Nations Conference on Trade and Development (UNCTAD) and Penn World Table (PWT) 7.0; Alan Heston, Robert Summers and Bettina Aten, Penn World Vision 7.0, Center for International Comparisons of Production, Income and Prices at the University of Pennsylvania, May 2011

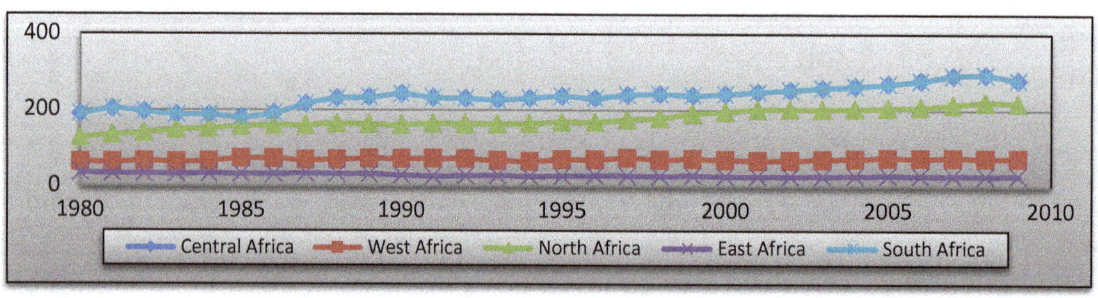

Graph 10.3 Manufacturing sector value added by region. Sources: Calculations made by the author based on data derived from the United Nations Conference on Trade and Development (UNCTAD) and Penn World Table (PWT) 7.0; Alan Heston, Robert Summers and Bettina Aten, Penn World Vision 7.0, Center for International Comparisons of Production, Income and Prices at the University of Pennsylvania, May 2011

When the curves (Graphs 10.4, 10.5, 10.6, 10.7, and 10.8 presented in Annexes) representing value added per capita in the regions that make up our sample are spontaneously analysed, two regions stand out from the rest. These regions are clearly responsible for the increase in the manufacturing average during the period considered in the analysis. These regions are Southern Africa (Graph 10.4) and North Africa (Graph 10.5). Since population size is considered, if we limit ourselves to the conclusions of development theories, it may be argued that the productivity rate in the manufacturing sector in these two regions is higher than the population growth rate. The other three regions, namely West, Central and East Africa are at the bottom of the scale. Each of them has barely one-third of the per capita value added generated by the manufacturing industry in North Africa. Rather than limiting ourselves to these initial findings concerning the different regions, we went on to compute and compare the growth rates of the manufacturing sector with those of the agricultural sector and the population (Table 10.3).

The analysis in Table 10.3 shows that in Africa the average growth rate of manufacturing sectors is lower than that of the agricultural sector and the

Table 10.3 Comparison between the manufacturing and agricultural sectors and population size in Africa

Indicators	Africa	North	South	West	Centre	East
Average level of agriculture	2603	5182	1325	2504	1551	3640
Average level of manufacturing sector	2118	5134	5137	564	770	857
Average population size (in thousand)	18,075	25,147	12,761	15,233	15,475	29,836
Average agricultural sector growth rate	6.0%	8.5%	5.0%	5.5%	5.0%	7.9%
Average manufacturing sector growth rate	3.4%	10.3%	8.2%	−3.1%	3.4%	6.1%
Average population growth rate	7.0%	5.6%	6.6%	7.3%	7.8%	7.3%
Difference between the manufacturing sector growth rate and the population growth rate	−3.7%	4.7%	1.6%	−10.4%	−4.4%	−1.2%

Sources: Calculations made by the author based on data derived from the United Nations Conference on Trade and Development (UNCTAD) and Penn World Table (PWT) 7.0; Alan Heston, Robert Summers and Bettina Aten, Penn World Vision 7.0, Center for International Comparisons of Production, Income and Prices at the University of Pennsylvania, May 2011

population. Their trends are such that the gap between the growth rate of manufacturing sector value added can be equated with that of the productivity of the sector, while the population growth rate of the entire continent is negative (−3.7%). The population and the agricultural sector value added are increasing twice as fast as the manufacturing sector value added. The gap between manufacturing sector and population growth rates is also negative in West Africa (−10.4%), Central Africa (−4.4%) and East Africa (−1.2%) but positive with respect to the population growth rates in North Africa (4.7%) and Southern Africa (1.6%). The differences in the manufacturing and agricultural sector growth rates are positive in North Africa and Southern Africa, and negative in Central, East and West Africa. The average growth rate of manufacturing sector value added for all the regions is positive. West Africa is the only region where it is negative (−3.1%). North, East and West Africa have the largest agricultural sectors. Central Africa and Southern Africa have the smallest sectors. Consequently, we cannot consider that the population growth rate and the size of the agricultural sector are the main factors responsible for Africa's average manufacturing performance.

We used the medians and averages of manufacturing value added by region to determine differences. Their closeness implies that differences in the manufacturing sector performance of the countries of the zone are small, or even almost nil. However, where there is a gap, it could be due to differences in the creation of manufacturing value added in the countries of the region.

In most regions, the median values of agricultural value added are fairly close to the average values of the variable. The same is true for the value added of the manufacturing sector, with the exception of Southern Africa where there are wide differences.

To carry out a more in-depth statistical analysis at the regional level, we opted to differentiate between countries based on their performance. There are three types of countries. Countries with the highest median manufacturing sector performance are considered as leading countries. Those with the highest median agricultural sector performance are referred to as countries with great potential because they can develop a viable agri-food industry though with a fairly low level of manufacturing sector value added (below the regional median). A country is considered to have performed well when its median value added is higher than the regional median of the group to which it belongs. Countries that cannot be classified under one of these two groups are those that need agricultural sector capacity building.

Based on this statistical analysis framework (Table 10.1), South Africa, Namibia and Botswana are regional industrial leaders in Southern Africa. Zambia is the only country in the region that has a great potential but does not seem to take advantage of it. In this group, Malawi, Mozambique and Zimbabwe need to be strengthened.

In North Africa, Morocco, Libya and Tunisia are leading countries, while Algeria and Egypt need to be strengthened.

For the other three regions, countries whose agricultural and/or manufacturing value added is close to that of those countries identified in the first two groups will be considered as leading countries or countries with great potential. We are obliged to use this approach because in these regions (West, Central and East Africa), the level of manufacturing value added per capita is much lower. The manufacturing median per capita in these three regions is about 80% less than that of North Africa and Southern Africa.

In East Africa where the agricultural median per capita is higher than that of Southern Africa, Kenya, Somalia and Sudan can be said to have an agro-industrial potential. Tanzania's agricultural sector needs to be strengthened. In Central Africa, Angola, Gabon and Equatorial Guinea have enormous potential. The agricultural and manufacturing sectors in Cameroon, which is the most industrially advanced country in the region, still need to be strengthened. The Central African Republic's agricultural sector in particular needs to be strengthened. In West Africa, Côte d'Ivoire is the only country with a level of manufacturing per capita close to that of the regional leaders identified. Although it is the regional leader, it has great potential like Ghana, Guinea-Bissau and Liberia. Benin, Senegal, Sierra Leone and Togo need agricultural capacity building. Overall, though North Africa and Southern Africa are leaders in the manufacturing sector, North Africa has the highest agricultural production level and the most stable group movement (in each of the sectors studied, the median is very close to the average). From the industrial viewpoint, the levels in Southern Africa are more diverse. For example, South Africa has a value of USD 808 per capita, while Botswana has a level of USD 175 per capita.

Structurally, the analysis in Table 10.4 shows that all the economic regions studied have a predominantly agricultural economic structure.

Table 10.4 Description of the structure of regional economies

Regions	Share of manufacturing sector in GDP (%)	Share of agricultural sector in GDP (%)	Per capita investment in USD	Human capital	Economic openness	Governance quality
Africa	1.12	24.33	256	42.88	60.57	0.41
North Africa	0.66	9.94	368	50.65	67.78	0.51
Southern Africa	1.63	14.02	562	71.88	64.32	0.49
Central Africa	0.96	16.51	70	35	31.68	0.37
East Africa	0.19	35.15	453	42.25	73.91	0.35
West Africa	1.49	34.8%	73	30.97	61.05	0.37

Sources: Calculations made by the author based on data derived from the United Nations Conference on Trade and Development (UNCTAD) and Penn World Table (PWT) 7.0; Alan Heston, Robert Summers and Bettina Aten, Penn World Vision 7.0, Center for International Comparisons of Production, Income and Prices at the University of Pennsylvania, May 2011

The share of manufacturing sector value added in gross domestic product (GDP) in North Africa is less than that in the West, Central and Southern Africa. East Africa has the lowest level on the continent. The only regions whose share of manufacturing sector value added in GDP is above the continental average (1.12%) are Southern Africa (1.63%) and West Africa (1.49%). The share of manufacturing sector value added in GDP of the other three regions, namely North, Central and East Africa, is below this average. Besides boosting productivity in various sectors, the objective of promoting structural change has not been achieved. Our approach is therefore well adapted to all the regions included in our study sample.

The analysis in Table 10.5 also shows that there is a positive correlation between the manufacturing value added and control variables, namely levels of investment per capita, human capital, degree of economic openness and governance quality. There is also a positive correlation between all variables and the agricultural sector value added.

10.5 Findings

It should be recalled that agricultural sector value added is the important variable in this study and that investment, human capital, economic openness and governance quality have been considered in our regressions. The different results obtained using fixed effects models, twin effects models (with temporal dummies) and the generalized least-squares estimator are very similar. The results of regressions using fixed effects models and random effects models for Southern Africa and North Africa (Table 10.6), Central Africa, East Africa and West Africa (Table 10.7) and the dynamic regression panel model using the generalized least-squares estimator (Table 10.8) are summarized in Table 10.9.

This section first presents a number of implications related to the links that exist between the two sectors under study and then those related to the control variables used.

The analysis shows that some regions of Africa have a great potential for achieving industrial development through primary products processing or

Table 10.5 Correlations

	Manufacturing value added per capita	Agricultural value added per capita	Investment per capita	Human capital	Economic openness	Governance quality
Manufacturing value added per capita	1.0000					
Agricultural value added per capita	0.3844	1.0000				
Investment per capita	0.5523	0.4890	1.0000			
Human capital	0.7092	0.4253	0.5943	1.0000		
Economic openness	0.1788	0.2149	0.4004	0.3221	1.0000	
Governance quality	0.4249	0.1184	0.2873	0.3703	0.1587	1.0000

Sources: Calculations made by the author based on data derived from the United Nations Conference on Trade and Development (UNCTAD) and Penn World Table (PWT) 7.0; Alan Heston, Robert Summers and Bettina Aten, Penn World Vision 7.0, Center for International Comparisons of Production, Income and Prices at the University of Pennsylvania, May 2011

Table 10.6 Results of the regressions for Africa, North Africa and Southern Africa

Regions	Africa			North Africa			Southern Africa		
Methods	Fixed effects models		Twin fixed effects models	Fixed effects models		Twin fixed effects models	Fixed effects models		Twin fixed effects models
	(1)	(2)	(3)	(1)	(2)	(3)	(1)	(2)	(3)
Variables	ln MVAPC	ln MVAPC	ln MVAPC	ln MVAPC	ln MVAPC	ln MVAPC	ln MVAPC	ln MVAPC	ln MVAPC
L.lnagrvapc	0.864**	0.999***	0.989***	−0.599	−0.514	−0.729	0.463	0.135	0.175
	(0.011)	(0.003)	(0.004)	(0.347)	(0.374)	(0.254)	(0.186)	(0.708)	(0.524)
L.haiwfg	0.006	0.006	0.010	0.019*	0.017*	0.036	0.015	0.014	0.023
	(0.200)	(0.177)	(0.285)	(0.070)	(0.094)	(0.102)	(0.132)	(0.153)	(0.154)
L.lngfcfpc	0.356***	0.310***	0.303***	0.518*	0.487*	0.501*	0.254	0.355*	0.444**
	(0.001)	(0.001)	(0.001)	(0.099)	(0.082)	(0.073)	(0.161)	(0.069)	(0.035)
L.openc	−0.002	−0.001	−0.001	−0.002	−0.001	−0.000	−0.003	0.000	0.001
	(0.205)	(0.527)	(0.676)	(0.532)	(0.694)	(0.925)	(0.318)	(0.768)	(0.435)
qogov		0.240	0.199		0.381	0.466		0.292	0.257
		(0.272)	(0.438)		(0.291)	(0.213)		(0.492)	(0.461)
3. Period			0.129**			−0.116			0.168**
			(0.034)			(0.195)			(0.046)
4. Period			0.066			−0.131			0.200*
			(0.380)			(0.298)			(0.078)
5. Period			−0.055			−0.258			0.014
			(0.527)			(0.266)			(0.941)
6. Period			0.016			−0.380			0.002
			(0.879)			(0.251)			(0.990)

7. Period			−0.057			−0.392			0.011
			(0.585)			(0.320)			(0.948)
8. Period			−0.031			−0.479			−0.084
			(0.823)			(0.301)			(0.660)
9. Period			−0.008			−0.526			−0.096
			(0.959)			(0.310)			(0.687)
10. Period			−0.026			−0.493			−0.160
			(0.889)			(0.418)			(0.606)
Constant	3.976***	4.266***	4.119***	0.168	0.420	−1.108	3.844**	2.308*	1.481
	(0.000)	(0.000)	(0.000)	(0.950)	(0.868)	(0.614)	(0.016)	(0.061)	(0.236)
Observations	331	324	324	43	43	43	63	61	61
R-squared	0.448	0.507	0.528	0.588	0.605	0.647	0.369	0.423	0.533
Number of id	37	37	37	5	5	5	7	7	7

Robust *p* value in parentheses

*** $p<0.01$, ** $p<0.05$, * $p<0.1$

Table 10.7 Results of the regressions for East Africa, Central Africa and West Africa

Regions	East Africa			Central Africa			West Africa		
Methods	Fixed effects models		Twin fixed effects models	Fixed effects models		Twin fixed effects models	Fixed effects models		Twin fixed effects models
	(1)	(2)	(3)	(1)	(2)	(3)	(1)	(2)	(3)
Variables	ln MVAPC	ln MVAPC	ln MVAPC	ln MVAPC	ln MVAPC	ln MVAPC	ln MVAPC	ln MVAPC	ln MVAPC
L.lnagrvapc	−0.320	0.398	0.405***	1.049**	1.064**	1.152*	1.434***	1.521***	1.527***
	(0.321)	(0.115)	(0.002)	(0.018)	(0.022)	(0.051)	(0.001)	(0.001)	(0.001)
L.haiwfg	0.009	0.021***	0.010*	0.031*	0.031*	0.020	−0.010*	−0.011	−0.016
	(0.222)	(0.001)	(0.079)	(0.067)	(0.063)	(0.514)	(0.097)	(0.136)	(0.583)
L.lngfcfpc	0.535***	0.147*	0.150***	0.520**	0.493*	0.542	0.218**	0.179**	0.191**
	(0.001)	(0.081)	(0.002)	(0.046)	(0.089)	(0.153)	(0.010)	(0.031)	(0.047)
L.openc	−0.002	−0.005*	−0.008***	−0.002	−0.002	−0.004	0.002	0.002	0.002
	(0.714)	(0.079)	(0.010)	(0.605)	(0.699)	(0.401)	(0.494)	(0.439)	(0.480)
qogov		0.640**	0.863***		0.399	0.579		0.348	0.351
		(0.046)	(0.000)		(0.535)	(0.465)		(0.395)	(0.519)
3. Period			−0.010			−0.137			0.189
			(0.850)			(0.122)			(0.110)
4. Period			−0.020			−0.141			0.087
			(0.429)			(0.492)			(0.636)
5. Period			−0.081			−0.282			0.075
			(0.154)			(0.331)			(0.663)
6. Period			0.005			−0.041			0.160
			(0.943)			(0.899)			(0.502)
7. Period			0.118**			0.054			−0.055
			(0.019)			(0.859)			(0.853)

8. Period			0.077			0.077			0.107
			(0.310)			(0.836)			(0.798)
9. Period			0.114			0.116			0.162
			(0.146)			(0.796)			(0.735)
10. Period			0.240**			0.118			0.226
			(0.019)			(0.823)			(0.722)
Constant	−0.130	2.373**	2.715***	2.449	2.463	2.987*	5.588***	5.799***	5.809***
	(0.868)	(0.034)	(0.000)	(0.111)	(0.113)	(0.085)	(0.000)	(0.000)	(0.000)
Observations	45	42	42	54	54	54	117	115	115
R-squared	0.486	0.828	0.934	0.728	0.733	0.787	0.610	0.631	0.663
Number of id	5	5	5	6	6	6	13	13	13

Robust *p* value in parentheses

*** $p<0.01$, ** $p<0.05$, * $p<0.1$

Table 10.8 Dynamic regression panel model

Region	Africa
Methods	Two-step generalized least-squares estimator
	(1)
Variables	ln MVAPC
L.lnMVAPC	0.516*** (0.000)
lnagrvapc	0.696*** (0.000)
haiwfg	−0.005 (0.645)
lngfcfpc	0.087** (0.049)
openc	0.003 (0.106)
qogov	0.141 (0.523)
2bn. Period	0.424 (0.150)
3. Period	0.331 (0.223)
4. Period	0.379 (0.152)
5. Period	0.211 (0.486)
6. Period	0.146 (0.561)
7. Period	0.771*** (0.002)
8. Period	0.297* (0.073)
9. Period	0.308* (0.059)
Observations	288
Number of id	37
Arellano-Bond test for AR(1)	−2.88* (0.04)
Arellano-Bond test for AR(2)	−1.48 (0.14)
Sargan test of overidentification restrictions	15.59 (0.96)
Hansen test of overidentification	14.86 (0.971)
Number of instruments	41

p value in parentheses
*** $p<0.01$, ** $p<0.05$, * $p<0.1$

Table 10.9 Results and estimates

Variables	Africa (Bulk Sample)	North Africa	Southern Africa	Central Africa	West Africa	East Africa
Agricultural sector value added	0.516***	(Ø)	(Ø)	1.152*	1.527***	0.405***
Investment	0.087**	0.501*	0.444**	0.493*	0.191**	0.150***
Human capital	(Ø)	(Ø)	(Ø)	(Ø)	(Ø)	0.010*
Economic openness	(Ø)	(Ø)	(Ø)	(Ø)	(Ø)	−0.008***
Governance quality	(Ø)	(Ø)	(Ø)	(Ø)	(Ø)	0.863***

* $p<0.1$, ** $p<0.05$, *** $p<0.01$

use of another form of agricultural surplus. We have demonstrated that there is a positive statistical correlation between agricultural and manufacturing sector value added. We have also underscored the positive impact of the agricultural sector on the manufacturing sector in Africa as well as in some regions. If the African continent seeks to achieve industrial development by following the path we have proposed, it should establish an entity responsible for implementing an appropriate industrialization policy and designing industrial policy implementation support and monitoring tools. Thus, it could be in a position to support national and regional economies in formulating their industrial policies and developing tools to implement and monitor such policies, taking into account the conditions and structure of each economy.

Despite some similarities, potential and, hence, results differ according to region. It was observed that some regions such as North Africa and Southern Africa have the best manufacturing, agricultural, investment, human capital, economic openness and governance quality performance. It was also realized that such performance could not bring about change in the structure of the economies of these regions which are still dominated by the agricultural sector. In addition, the size of the agricultural sector does not seem to produce any impact on the manufacturing sector in these two regions which apparently produce mainly agricultural products for local consumption or export. Thus, to build agro-based industries, countries in these regions should first of all reorient the use of agricultural products, while making the necessary investments. These are merely options for improving manufacturing value added, given that the two regions have clearly opted for an industrialization model that differs from our structural change approach. They can, however, adopt the approach. As indicated earlier, these regions have the highest levels of manufacturing value added, but the manufacturing sector's share in the economy suggests that structural change is in progress.

The level of manufacturing in Central, West and East Africa, unlike in Southern Africa and North Africa, is low. West and Central Africa have the highest agricultural sector contribution to the manufacturing sector with a low total agricultural value added and the lowest manufacturing value added on the continent. The agricultural sector's contribution to the manufacturing sector in East Africa is positive, but low. The very small share of manufacturing value added in total value added suggests that these regions should make more investments in agricultural production and adopt measures to ease the transfer of agricultural surplus to the manufacturing sector. This weakness may highlight the lack of an industrial policy or inadequate application thereof to develop the agri-food industry.

The analysis of these results shows that it is necessary to improve human capital in Africa in order to promote change in the continent's economic structure and in the various regions. This reflects the need to provide appropriate training and disseminate knowledge to foster knowledge absorption, improve productivity and promote structural change.

The absence or negative impacts of economic openness underscore the need to improve countries' trade policies and build trade capacity to achieve greater global trade integration. Trade with the outside world does not promote the development of the agri-food industry. Thus, it does not adequately promote technology and scientific knowledge dissemination. There is need to adopt a trade-opening policy that promotes structural change.

The absence of governance quality impacts makes it necessary to strengthen countries' institutional structures and tailor them to their needs. At any rate, it is necessary to improve governance quality to ensure the proper selection of investments and successful implementation of adapted economic policies.

In Central Africa, investment is making a significant contribution and should be sustained and promoted to enhance the development of the manufacturing sector. The agricultural products value added should also be improved to promote structural change.

The manufacturing sector's share in West African countries' GDP is one of the highest on the continent, but certainly one of the lowest among developing countries worldwide. The improvement of investments in both sectors will not only promote their development, but also improve the overall impact of investment, which is very low.

In East Africa, efforts should mainly focus on governance quality, which is the most important element of industrial development in the region.

As we have seen, West and Central Africa have much better results. East Africa can achieve similar results, but negative shocks, though temporary, may limit the impact of such results.

Although Southern Africa and North Africa have the highest level of manufacturing sector performance, they are no longer leaders in agricultural products processing or the mobilization of agricultural surplus. West and Central Africa have the highest contribution.

10.6 Conclusion

This study focused on an industrial development model based on agricultural production. First, we reviewed the works of Lewis (1954), Mellor J. W. (1966) and Kuznets (1973) to circumscribe the topic. This helped to better understand the link between the agricultural and manufacturing sectors as well as the mechanisms through which the agricultural sector promotes the development of manufacturing activities.

In the approach adopted, we followed in the footsteps of these preceding authors by underscoring the size of the agricultural sector as a prerequisite for industrial take-off. Next, global and regional statistical analyses showed that countries with the highest level of manufacturing activities in Africa also enjoy good agricultural performance. This group includes South Africa, Swaziland, Namibia, Morocco, Libya, Tunisia, Central African Republic, Cameroon, Gabon, Botswana, Côte d'Ivoire and Ghana as potential leaders of industrialization in Africa.

We obtained relevant results concerning industrialization based on agricultural products. The econometric regressions conducted suggest that it is not easy for all countries to adopt this development model. North Africa and Southern Africa have the highest levels of manufacturing value added. When the different factors that influence manufacturing value added in Africa are considered, it becomes quite clear that the size of the agricultural sector is not necessarily the element that promotes the industrial development of these regions. In addition, our results enabled us to identify the regions that seem to benefit most from the impacts of agricultural sector development on the manufacturing sector, namely West and Central Africa. Yet, these regions are among those with the lowest manufacturing value added in Africa. Despite this low value creation, we were able to reclassify the regions according to contribution, based on the value and significance of the coefficient associated with agricultural value added. According to this new classification, West Africa is followed by Central Africa and then by East Africa, whose performance is undermined by the negative effects of temporary shocks.

Southern Africa and North Africa are at the bottom of the ranking for the simple reason that they mainly produce for export or local consumption. This is reflected in their high levels of manufacturing and agricultural sector value added, but without any significant contribution to the first sector.

We were able to identify some of the factors responsible for the weakness of the agricultural sector-based industrial development model in Africa. These include low skill-development potential or level of human capital, the educational thrusts chosen, the educational and skills utilization policies, poor infrastructure, low level of investment and poor targeting of investments, the adoption of trade policies that do not promote this type of industrial development and the acute need for institutional capacity building that could certainly promote such structural change. These elements have enabled us to make some recommendations to improve manufacturing sector value added. These include the implementation of agricultural and industrial sector development support policies, the redefinition of educational policies, the development of transport infrastructure, more effective targeting of investments to be made, the adoption of a trade policy that promotes economic integration, the use of technology and efforts to improve governance quality.

Annexes: Agricultural and Manufacturing Sector Value Added by Region

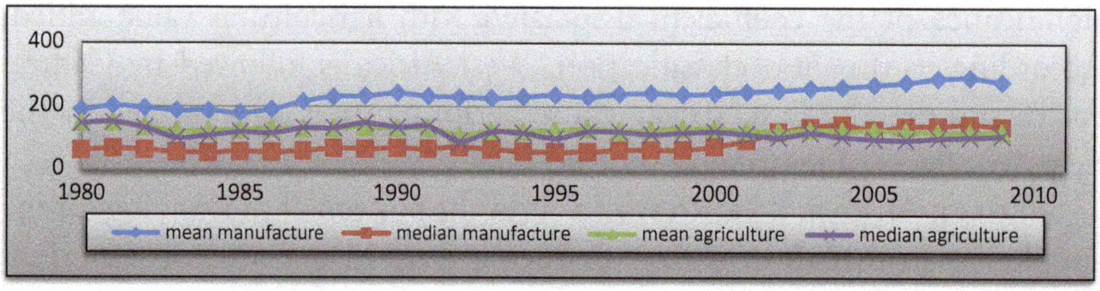

Graph 10.4 Southern Africa. Sources: Calculations made by the author based on data derived from the United Nations Conference on Trade and Development (UNCTAD) and Penn World Table (PWT) 7.0; Alan Heston, Robert Summers and Bettina Aten, Penn World Vision 7.0, Center for International Comparisons of Production, Income and Prices at the University of Pennsylvania, May 2011

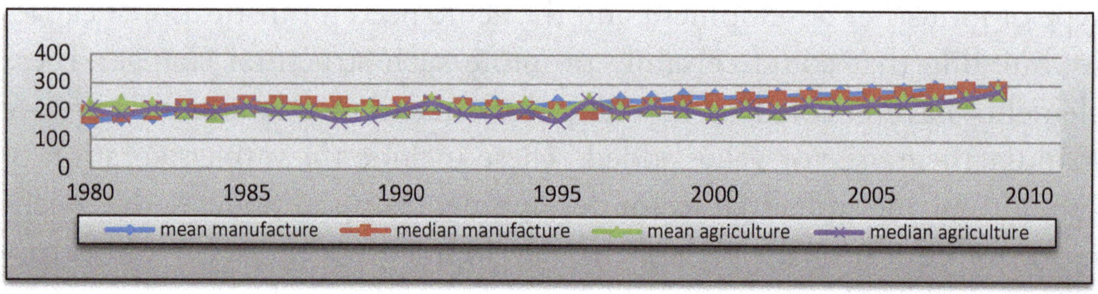

Graph 10.5 North Africa. Sources: Calculations made by the author based on data derived from the United Nations Conference on Trade and Development (UNCTAD) and Penn World Table (PWT) 7.0; Alan Heston, Robert Summers and Bettina Aten, Penn World Vision 7.0, Center for International Comparisons of Production, Income and Prices at the University of Pennsylvania, May 2011

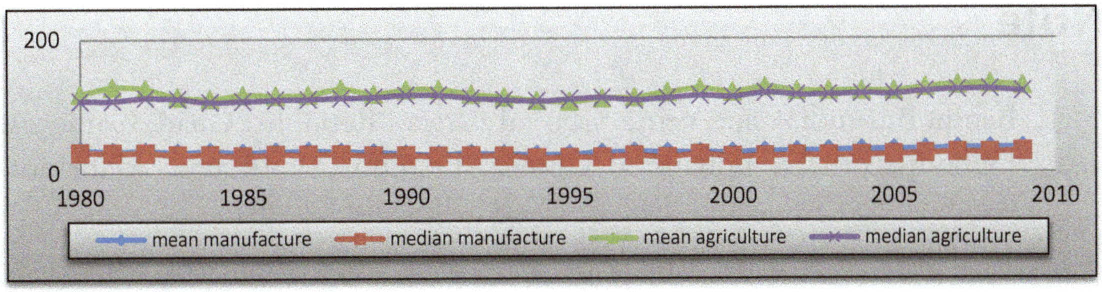

Graph 10.6 East Africa. Sources: Calculations made by the author based on data derived from the United Nations Conference on Trade and Development (UNCTAD) and Penn World Table (PWT) 7.0; Alan Heston, Robert Summers and Bettina Aten, Penn World Vision 7.0, Center for International Comparisons of Production, Income and Prices at the University of Pennsylvania, May 2011

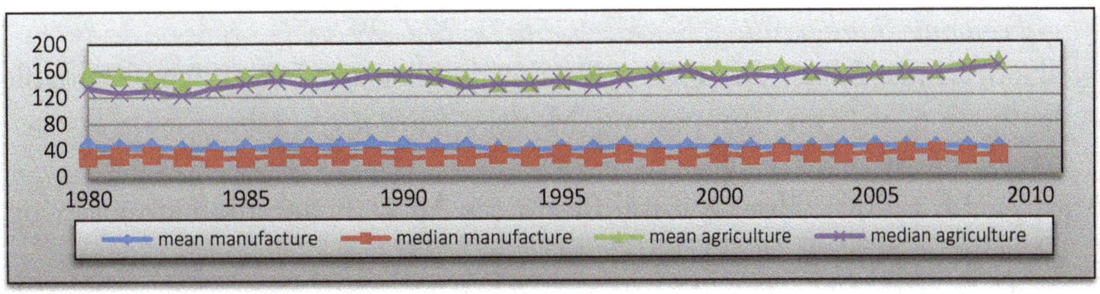

Graph 10.7 West Africa. Sources: Calculations made by the author based on data derived from the United Nations Conference on Trade and Development (UNCTAD) and Penn World Table (PWT) 7.0; Alan Heston, Robert Summers and Bettina Aten, Penn World Vision 7.0, Center for International Comparisons of Production, Income and Prices at the University of Pennsylvania, May 2011

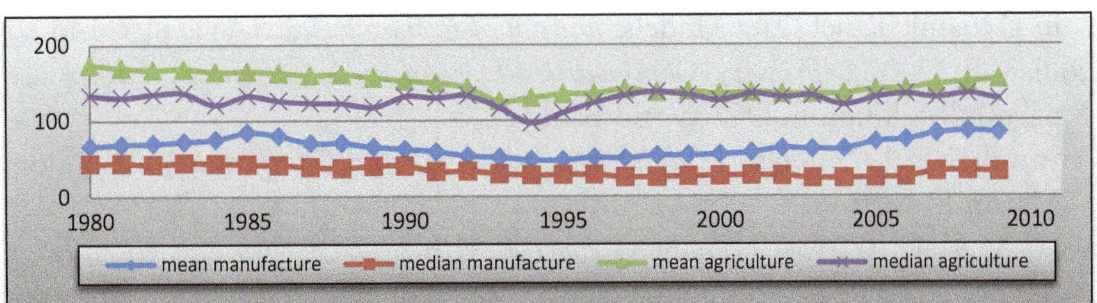

Graph 10.8 Central Africa. Sources: Calculations made by the author based on data derived from the United Nations Conference on Trade and Development (UNCTAD) and Penn World Table (PWT) 7.0; Alan Heston, Robert Summers and Bettina Aten, Penn World Vision 7.0, Center for International Comparisons of Production, Income and Prices at the University of Pennsylvania, May 2011

Note

1. Benin, Burundi, Cape Verde, Central African Republic, Chad, Comoros, Djibouti, Eritrea, Equatorial Guinea, Lesotho, Mauritania, Mauritius, Rwanda, Sao Tome and Principe, Seychelles and Swaziland.

References

Abdelmalki, L., & Mundler, P. (1995). *Economie du développement, les théories, les expériences et les perspectives* (311 pages). Paris: Hachette Supérieur.

Arellano, M., & Bond, S. (1991). Some Tests of Specification for Panel Data: Monte Carlo Evidence and an Application to Employment Equations. *Review of Economic Studies, Wiley Blackwell, 58*(2), 277–297.

Arellano, M., & Bover, O. (1995). Another Look at the Instrumental Variable Estimation of Error-Components Models. *Journal of Econometrics, Elsevier, 68*(1), 29–51.

Auty, R. M. (2000). How Natural Resources Affect Economic Development. *Development Policy, 18*, 347–364.

Bairoch, P. (1971). *Le Tiers-Monde dans l'impasse: le démarrage économique du XVIIIe au XXe siècle* (Vol. 250). Paris: Gallimard.

Balassa, B. (1979). The Changing Pattern of Comparative Advantage in Manufactured Goods. *The Review of Economics and Statistics, 61*(2), 259–226.

Barro, J. (1998). *Determinants of Economic Growth: A Cross-Country Empirical Study* (Vol. 1, 1st ed.). Cambridge, MA: The MIT Press.

Blundell, R., & Bond, S. (1998). Initial Conditions and Moment Restrictions in Dynamic Panel Data Models. *Journal of Econometrics, 87*(1), 115–143.

Collier, P. (2002). *Primary Commodities Dependence and Africa's Future* (26 pages). Washington, DC: World Bank.

Cornwall, J. (1977). *Modern Capitalism: Its Growth and Transformation*. London: Martin Robertson.

ECA, & AUC. (2013). *Economic Report on Africa 2013: "Making the Most of Africa's Commodities: Industrializing for Growth, Jobs and Economic Transformation"*. Addis Ababa: United Nations Publication, E.13.II.K.1.

Fei, J. C., & Ranis, G. (1969). Economic Development in Historical Perspective. *The American Economic Review, 59*, 386–400.

Fondation pour les Etudes et la Recherche sur le Développement International (FERDI). (2011). Human Asset Index Computing Retrospective Series from 1970–2008. 55 pages.

Hawkins, A. M. (1986). Can Africa Industrialize? In *Strategies for African Development* (pp. 279–307). Berkeley, CA: University of California Press.

Hughes, H. (1984). *Industrialization and Development: A Stocktaking. Industrialisation and Development. A Third World Perspective.* Westport: Greenwood Press.

Kuznets, S. (1966). *Modern Economics and Growth: Rate Structure and Spread.* New Haven and London: Yale University Press.

Kuznets, S. (1973). Modern Economic Growth: Findings and Reflections. *The American Economic Review, 63*, 247–258.

Lewis, W. A. (1954). Economic Development with Unlimited Supplies of Labour. *The Manchester School, 22*, 139–191.

Mayer J., & Fajarnes P. (2008). Tripling Africa Primary Exports: What? How? Where? UNCTAD Discussion Papers 191, United Nations Conference on Trade and Development, 39 pages.

Mellor, J. W. (1966). *The Economics of Agricultural Development.* Ithaca: Colonel University Press.

Park, B., & Lee, K.-K. (2006). Natural Resources, Governance, and Economic Growth in Africa. *Journal of International Economic Studies, 10*(2), 1598–2769.

Roodman, D. (2006). How to Do xtabond2: An Introduction to "Difference" and "System" GMM in Stata. *Stata Journal, StataCorp LP, 9*(1), 86–136, 53 pages.

Shifa, A. B. (2011). *Does Agricultural Growth Have a Causal Effect on Manufacturing Growth?* (13 pages). Stockholm: Institute for International Economic Studies, Stockholm University.

Studennund, A. H. (2000). *Using Econometrics: A Practical Guide* (4th ed.). Boston, MA: Addison-Wesley.

Szirmai A., & Verspagen B. (2011). Manufacturing and Economic Growth in Developing Countries, 1950–2005, wp2011-069, Maastricht Economic and Social Research and training Centre on Innovation and Technology, United Nations University (UNU-MERIT) and Maastricht University, The Netherlands, 41 pages.

Tregenna, F. (2007). *Which Sectors Can Be Engines of Growth and Employment in South Africa? An Analysis of Manufacturing and Services.* UNU-WIDER Conference on 'Southern Engines of Global Growth: China, India, Brazil and South Africa', Helsinki, Finland, Vol. 7.

United Nations Conference for Trade and Development. (2013). Intra-African Trade: Unlocking Private Sector Dynamism, 158 pages.

United Nations Conference for Trade and Development Database. (2011). http://unctadstat.unctad.org/ReportFolders/reportFolders.aspx?sCS_referer=&sCS_ChosenLang=fr.

Wiggins V. (2003). http://www.stata.com/statalist/archive/2003-10/msg00031.html.

Wood, A., & Mayer, J. (2001). Africa's Export Structure in a Comparative Perspective. *Cambridge Journal of Economics, 25*(3), 369–394.

11

Transforming African Agriculture Through Special Economic Zones: Opportunities and Challenges

Joseph Tinarwo

11.1 Introduction

This chapter interrogates the effectiveness of the special economic zones (SEZs) in bringing about African agriculture transformation. The chapter opens with the presentation of the materials and methods used in its development, and this is followed by a section on the inclusiveness of agriculture for industrialization and economic growth. In this section, literature from various scholars is presented showing the importance of agricultural transformation in ensuring desirable socio-economic outcomes. The historical development of SEZs is traced; their typologies and also their merits and side effects based on practical implementation realities from the selected case studies. The last part of the chapter gives the recommendations for the successful implementation of SEZs for agricultural transformation in Africa.

African governments are under pressure to transform agriculture in order to fight the protracted food and nutrition insecurity and also meet both regional and global commitments such as Sustainable Development

J. Tinarwo (✉)
Great Zimbabwe University, Masvingo, Zimbabwe

A. Shimeles et al. (eds.), *Building a Resilient and Sustainable Agriculture in Sub-Saharan Africa*, https://doi.org/10.1007/978-3-319-76222-7_11

Goals (SDGs), Comprehensive Africa Agriculture Development Programme (CAADP) among others. Notwithstanding the significant progress in transforming agriculture, the continent remains the net food importer and has experienced an increase in the number of undernourished people over the past 30 years. In effect, sub-Saharan Africa (SSA) is the region with the highest prevalence of hunger with one person in four being undernourished (FAO 2015). It is axiomatic that the world's population is growing and Africa's population is expected to grow the fastest with UN Department of Economic and Social Affairs report of 2015 estimating that Africa will have two billion people by 2050. FAO (2015) cautions that the demand for food is expected to grow due to population growth, thus calling for stronger interventions to arrest the situation and finally eliminate hunger, achieve food security and improve nutrition, and promote sustainable agriculture. Agricultural transformation should become a top priority since agriculture is the backbone of African economies accounting to over 30% of the gross domestic product (GDP) for many African countries and remains the primary activity of over 60% of the African population (ACBF 2012: iv; AfDB 2016: 1). Regardless of the fact that agriculture accommodates the prime share of most African economies and supports both rural and urban livelihoods, it still endures a horde of challenges. To this end, the transformation of agriculture is imperative and SEZs are one of the vehicles that can position African Agriculture on a growth trajectory. Empirical evidence suggests that successful implementation of SEZs in agriculture results in employment creation, GDP growth, improved standards of living, technology, and industrial development.

11.2 Materials and Methods

This chapter is basically a reflection on the resources obtained from various sources such as the World Bank (WB), United Nations Development Programme (UNDP), Food and Agriculture Organization (FAO), International Labour Organization (ILO), African Development Bank (AfDB), and African Capacity Building Foundation (ACBF) among other development organizations. This was cemented by various reports from agricultural research organizations coupled with the reputable journal publications. Key informants' interviews were also done with officials working with the Ministry of Macro-Economic Planning and Investment Promotion in Zimbabwe and also economic experts of other government ministries, University of Zimbabwe, and Great Zimbabwe University on the concept of SEZs. Both the descriptive and comparative methods together with thematic analysis were used in developing this chapter.

11.3 Agricultural Transformation for Inclusive Growth

While the role of agriculture in economic growth and structural transformation is widely acknowledged, its characteristics in recent years have created a daunting task for policymakers in order to realize its gains. To Olaoye (2014), agriculture plays a critical role in the socio-economic activities of any given country. The WB (2007) reinforces that agriculture directly contributes to economic growth and enhances growth in other sectors through consumption and production linkages with agro processing and food marketing boosted, while backward linkages increase demand for immediate inputs and services. At the 2009 World Food Summit, the heads of governments unanimously agreed that poor countries needed to develop economic and policy tools to boost their agricultural production and competitiveness. Furthermore, a call for an increased agricultural investment was made at this summit since for majority of poor countries a vibrant agricultural sector is essential to overcome hunger and poverty. In fact, the agricultural sector is a prerequisite for the overall economic growth for most African countries. Olaoye (2007) took the argument further indicating that agriculture can enhance an increase in GDP, provide food and employment for the people and thus reduce poverty. In light of the African Union's (AU) Vision 2063, accelerating industrialization is a critical cog for African countries to reduce poverty and achieve economic growth (UNDP 2015: 10). Therefore, SEZs are an imperative route that African governments can utilize in overcoming the constraints of scale and competitiveness.

In essence, SEZs foster the creation of an enabling business environment, improved policies, infrastructure, and competitive transaction hence resulting in enhanced agricultural transformation.

11.4 Historical Developments of Special Economic Zones

The development of modern SEZs can be traced as far back as the 1950s in Ireland. SEZs then later spread to Latin America and East Asia in the 1970s under the various formulation and sectorial focus with governments trying to find pathways to industrialization (Baissac 2011). Today, SEZs are now a common global feature to solve many economic woes with most African countries embracing them by following the successful Chinese model. From a few dozens in the 1950s, today, the number of SEZs has ballooned to more than 3000 as instruments for the industrialization process, especially as a way of attracting foreign direct investments (FDIs), creating jobs, and generating exports and foreign exchanges (Zeng 2015: 3).

11.5 Typologies and Scope of Special Economic Zones

Though literature is awash with definitions of SEZs, Baissac (2011) coined that SEZs refer to a policy concentrate designed to increase growth by creating an economic environment which offers significantly better investment and operating conditions than the rest of the domestic economy, and ensures that conditions of international competitiveness are created. SEZ refers to a geographical region that has economic laws that are more liberal than a country's typical economic laws, and in many cases, it offers high-quality infrastructure facilities and support services and allows duty-free imports of capital goods and raw materials (Singh 2013; Farole 2011). The key characteristics of SEZs according to the World Bank (2008) include the following: (a) a physically secured and demarcated geographical area, (b) a single management or administration, (c) offers benefits for investors physically within the zone, and (d) streamlined procedures with duty-free benefits. Inherently, SEZs differ in terms of types, objectives, markets, and activities. Zeng (2015: 3) argues that SEZs manifest in an extensive array of forms including free-trade zones, export processing zones, industrial parks, economic and technology development zones, high-tech zones, science and innovation parks, and free ports, among other enterprise zones. Intrinsically, World Bank's Foreign Investment Advisory Service (FIAS) now (Investiment Climate Assessments (ICAS)) did a comprehensive mapping on the performance and typologies of SEZs in 2008 with the common ones being enterprise zones, free enterprises, free trade zones, free ports and export processing zones (FIAS 2008).

11.5.1 Free-Trade Zones

Free-trade zones, also known as commercial free zones, are small, fenced-in, duty-free areas, offering warehousing, storage, and distribution facilities for trade, transshipment, and re-export operations and are usually located in most ports of entry around the world. In fact, free-trade zones are the most ubiquitous and oldest form of SEZs and a famous example is the Colon Free Trade Zone in Panama.

11.5.2 Export Processing Zones

This type of SEZs can be traced as far back as 1950s and were initially implemented in South Korea and Ireland. Export processing zones (EPZs) aim at accelerating industrialization mostly for export markets and classically take two forms. In the traditional EPZ model, the entire area within the zone is exclusively for export-oriented enterprises licensed under an EPZ regime. Hybrid EPZs, in contrast, are typically subdivided into a general zone open to all industries regardless of export orientation and a separate EPZ area reserved for export-oriented, EPZ-registered enterprises.

11.5.3 Free Ports

These are commonly broader and classically encompass much larger areas and may include both urban and rural territories. This type of SEZs incorporate large transport facilities like ports, airports, and goods and services-related trade activities and a good example is the large-scale free ports in China. Free ports thus incorporate entire economic regions, the population that live and work in these regions, and the entirety of the economic activities that take place there.

11.5.4 Free Enterprises

These are also called single-company zones and are a variation of the EPZs, where the EPZ status is afforded to single enterprises outside the zone. Implied in this type of SEZs is that it provides incentives to individual enterprises regardless of location; factories do not have to locate within a designated zone to receive incentives and privileges. Primary examples of countries relying exclusively on a single factory scheme include Mauritius, Madagascar, Mexico, and Fiji; other countries such as Costa Rica, the United States, and Sri Lanka allow both industrial estate-style zones and single factory designations.

11.5.5 Enterprise Zones

Enterprise zones are a type of SEZs meant for economic revitalization of distressed urban or rural areas through the provision of tax incentives and financial grants. This type of zone is in developed countries, for example, the United States, France, and the United Kingdom, although South Africa is developing a similar mechanism. In effect, enterprise zones in these case studies have sought to bring regeneration and economic diversification to once striving regions.

11.6 Virtues and Potential Side Effects of Special Economic Zones in Agricultural Transformation

Despite the increased rhetoric and enthusiasm for SEZs in recent years, the practical implementation realities indicate that they bring about mixed results. Zeng (2015) applauded SEZs as highly effective tools for job creation. Empirical evidence suggests that SEZs are more significant sources of employment in smaller countries with populations less than five million such as Mauritius, Seychelles, and Jamaica than in large countries (FIAS 2008: 3). Zeng (2015: 3) reinforces that the popularity of SEZs is registered by two main benefits: that is, static economic benefits which include employment creation, export growth, increase in government revenues, and foreign exchange earnings, while the broader dynamic economic benefits include skills upgrading, technology transfer, economic diversification and innovation, and productivity enhancement of local firms.

They are capable of contributing to export development in terms of both accelerating export growth and diversification, and this is particularly important for poor developing countries reliant on export of primary products. According to Export Promotion Council for SEZs of India, SEZs exports accounted for 26% of India's total exports in the year 2011 with the Ministry of Industry and Commerce arguing that between 2013 and 2014 total exports from SEZs generated USD 82.35 billion. Moreover, SEZs can be instrumental in attracting FDI, offsetting some aspects of an adverse investment climate by offering worldwide class and best practice policies. UNDP (2015: 10) vows that African SEZs offer a number of advantages to investors, such as reduced customs duties and value-added taxes; simplification and centralization of administrative procedures through "one-stop-shops"; access to key national and international infrastructure; secured access to and reduced factor costs for electricity, water, and telecommunication services; relaxation of foreign exchange regulations; preferential interest rates offered by local banks; and reduced freight rates. In return, African governments are putting regulations in place that oblige investors to create local unskilled and skilled jobs, ensure linkages with the local economy and transfer technology and knowledge, while complying with local social and environmental regulations.

Despite the virtues of SEZs, there are some potential side effects that African governments can avoid in pursuit of the agricultural transformation agenda through SEZs. Zeng (2015: 7) highlights that SEZs may result in environmental degradation, for example, in China, the GDP performance used to be the top priority for the government officials without looking at the effects of SEZs' implantation process to the environment. The WB estimates that the environmental costs are about 8% of GDP and, to address this, China has since implemented tougher environmental standards and tried to use fiscal policies to force firms to adopt "green technologies" and conduct innovations (ibid.). ILO (2012) cautions that in some countries, SEZs have been castigated for deleterious socio-economic results especially to the women, youth, and working environments. Some of the SEZs probable pitfalls include labor exploitation especially among women and youth coupled with low wages, inadequate training and skill upgrading, use of trainees to lower wage costs, subdual of labor rights, and lax environmental standards (ibid.). Singh (2009) observes that if SEZs are set up on agricultural land, they create obstacles for the social and economic development of the country, especially if fertile land area under agriculture is acquired. A notable case is India's Singur and Nandigram where the government acquired land forcibly from the farmers at lower prices and gave SEZ developers at a subsidized rate, thus resulting in farmer agitations against the government (ibid.).

11.7 Global and Regional Experiences of Special Economic Zones

11.7.1 China

China is regarded as a global classic case in the successful development and implementation of SEZs with the country recording the leading destination of FDI in the developing world. Baissac (2011) advances that

China records more than 200 SEZs of various types, sizes, and industrial focus and has started expanding the model to other developing countries of the world including Africa. Shenzhen is one of the cities that were transformed by SEZs from agriculture-based economy in the 1970s to an industry-based economy in the early 1980s. Recently, Shenzhen is regarded as one of China's mainland cities in terms of economic returns amounting to USD 27.88 billion in local revenue in 2013, up 16.8% from 2012. Inherently, based on the overall statistics obtained from China Development Bank in May 2015, the contributions of SEZs to technological progress and innovation in the agricultural sector stands at 55.2%, while in agro-tech parks and agricultural demonstration zones, the contribution rate of technology reaches roughly 70%, nearly the average level of developed nations. In addition, these parks have also significantly contributed to the increase of farmers' income—on average, agricultural incomes within these parks are over 30% higher than incomes in surrounding villages, (Zeng 2015: 5).

11.7.2 India

SEZs are seen as engines of economic growth in India, and they play a vital role in the country's export strategy. Ideally, SEZs in India existed before the promulgation of the SEZ Act in 2005, which became operational in 2006 (Dohrmann 2008). This piece of legislation aims to give an all-inclusive policy framework to the key players in the SEZ program. SEZs in India aim at promoting industrialization and economic growth through tax rebates, fiscal incentives, and lands at subsidized rates. Agriculture-related SEZs in India include Falta food processing unit at West Bengal and Hassan with an area of 157.91 hectares. Despite all odds, exports through Indian SEZs grew further by 15.4% to reach USD 66 billion. As at 2011–2012 fiscal year, investments worth over USD 36.5 billion have been made in these tax-free enclaves. Exports of Indian SEZs have experienced a phenomenal growth of 50.5% for the past eight fiscals from a meager USD 2.5 billion in 2003–2004 to about USD 65 billion in 2011–2012 (accounting for 23% of India's total exports). Despite the benefits enjoyed by the Indian government through the SEZs, there are issues that need to be addressed, especially from the indigenous smallholder farmers who are losing their agricultural productive land to pave way for the establishment of SEZs, thus putting their food security and livelihoods at stake. It is thus imperative for the government of India to make sure that land acquisition and SEZs must prove beneficial for the local people.

11.7.3 Mauritius

The Government of Mauritius is considered as one of the success stories of SEZs implementation in Africa. Since the beginning of thc 1970s, thc Mauritius government has been committed in the SEZ development process by creating an enabling environment for investment and attracting FDI. Since their inception, the government has implemented two models of SEZs as turnaround strategies for their economy, that is, export processing zones and free port zones. The Mauritius government signed agreements to attract investors from China through joint ventures with local companies resulting in export-led growth and quick knowledge transfer (UNDP 2015: 11). In 2006, the government introduced the Business Facilitation Act to give more incentives to players in all sectors of the economy, and today, the whole country is regarded as an SEZ with the highest ease of doing business profile in the region according to the 2016 WB ranking.

11.7.4 Mozambique

In a bid to restore macroeconomic stability after the 1992 peace agreement, Mozambique's experience with SEZs is relatively recent evidenced by the Government Decree no. 75/2007, which established the Office for Accelerated Economic Development Zones. Kirk (2014: ix) informs that in 1998 additional specific incentives were introduced for SEZs and Industrial Free Zones in 1999 following the Chinese model with comprehensive fiscal benefits which included tax holidays, customs duty and indirect tax exemptions, along with tax credits. In 2015, about five SEZs have been established in the country with the other development corridors identified for the possible creation of SEZs for agriculture. According to the Minister of Agriculture and Food Safety (José Pacheco), agricultural development corridors were determined based on agro-climate conditions, strategic location vis-à-vis markets, existing or planned infrastructures, and the need to diversify farm products. He further highlights that opportunities in the value chain for products such as potatoes, wheat, beans, maize, soy, rice, and others deriving from poultry, cattle, and forestry activities have been identified in the six corridors (http://www.macauhub.com.mo/en).

11.7.5 Zimbabwe

Since the year 2000, Zimbabwe has experienced an economic downturn characterized by a hyperinflationary environment, low industrial capacity utilization, decline in agricultural production, and high imports to meet food and industrial raw materials among others (IISD 2009, http://www.afdb.org). The Government of Zimbabwe recently re-established SEZs in order to address the myriad of economic challenges, restore the productive sectors' status, and also improve exports through beneficiation and value addition which resonates with the country's blueprint, the Zimbabwe Agenda for Socio-Economic Transformation (ZIMASSET). The earmarked SEZ for agriculture targets Mashonaland Central and West Provinces focusing on food crop production and processing on such crops as maize, wheat, and soya beans, among others, with the Eastern Highlands specializing on fruit processing. Though the concept of SEZs is not entirely new in Zimbabwe, the government has once implemented EPZ initiative from 1996 to 2006 under the auspices of Export Processing Zones Act of 1995. The SEZs Bill was approved and signed in a law by the President on November 1, 2016, and this paved the way for the establishment of the SEZs Authority. Despite these developments, the establishment of SEZs had ushered mixed views from different stakeholders including labor unions especially clause 56 of the SEZs Bill which proposed to exempt investors licensed in these zones from the provisions of the Labor Act, as well as the Indigenization and Economic Empowerment Act which requires 51% stake from foreign investors. The Zimbabwean government has to create a favorable environment and ensure that there are sound policies in order to realize the gains that come with SEZs. This should be augmented by systematic research and dialogue among various stakeholders to ensure that SEZs have a transformational impact in Zimbabwe.

11.8 Conclusions and Recommendations

There is huge potential in developing agro-based industries for inclusive growth through SEZs and success stories from the implementation of SEZs for agriculture in China, India, Mauritius, and other countries should be a motivating factor for other African countries to embrace them. In essence, SEZs are instruments for the industrialization process, attracting FDIs, creating jobs, and generating exports and foreign exchanges that are critical in tapping the nascent agriculture in Africa. However, the mixed results of SEZs in many countries demonstrate that they are not an automatic antidote to the socio-economic challenges but rather have to be appropriately executed and tailored to suit the specific country context. Given the intricate and assorted contexts in which SEZs exist, it is essential to do research, study tours of successful SEZs, and establish the legal and instructional frameworks in order to guide their development and implementation. The findings from the 2015 comparative study by UNDP and International Poverty Reduction Center in China about African SEZs suggest the following recommendations in order to ensure SEZs' success in Africa:

- Ensure high-level political commitment and support for effective inter-ministerial collaboration
- Integrate SEZ programs into national development strategies and plans
- Support all industries that have a comparative advantage through SEZ development
- Ensure sufficient funding for infrastructure development within, and availability of good infrastructure outside, the SEZ prior to the SEZ approval

- Provide incentives for the creation of joint ventures between foreign SEZ companies and local companies
- Respond to SEZ labor requirements by aligning curricula of universities and Technical Vocational Education and Training (TVET) institutions
- Set high environmental standards in line with the United Nations Industrial Development Organization's Guidelines for Green Industry Parks and put a system in place to ensure their enforcement
- Establish low minimum SEZ investment thresholds for established local companies

Notwithstanding the imperative need to understand the effects of SEZs on agricultural transformation in Africa, the empirical data are still limited, thus calling for further research to inform policymakers and various stakeholders on the realities of their practical implementation.

References

ACBF. (2012). *Africa Capacity Indicators: Capacity Development for Agricultural Transformation and Food Security*. Harare: ACBF.

AfDB. (2016). *Feed Africa: Strategy for Agricultural Transformation in Africa 2016–2025*. AfDB, Abidjan 01, Côte d'Ivoire.

Baissac, C. (2011). Brief History of SEZs and Overview of Policy Debates. In T. Farole (Ed.), *Special Economic Zones in Africa: Comparing Performance and Learning from Global Experience*. Washington, DC: World Bank.

Dohrmann, J. A. (2008). Special Economic Zones in India – An Introduction, ASIEN 106.

Farole, T. (2011). *Special Economic Zones in Africa: Comparing Performance and Learning from Global Experience*. Washington, DC: World Bank.

FIAS. (2008). *Special Economic Zones: Performance, Lessons Learned, and Implications for Zone Development*. Washington, DC: World Bank.

FAO. (2015). *Regional Overview of Food Insecurity: African Food Security Prospects Brighter Than Ever*. Accra: FAO.

IISD. (2009). *Agriculture: Future Scenarios for Southern Africa – A Case Study of Zimbabwe's Food Security*. Winnipeg: International Institute for Sustainable Development (IISD).

Kirk, R. (2014). Special Economic Zones and Economic Transformation: An Assessment of the Impact of Special Economic Zones in Mozambique, USAID.

ILO. (2012). *Trade Unions and Special Economic Zones in India*. Geneva: ILO. Retrieved July 9, 2016, from http://www.ilo.org/wcmsp5/groups/public/---ed_dialogue/.

Olaoye, O. (2014). Potentials of the Agro Industry Towards Achieving Food Security in Nigeria and Other Sub-Saharan African Countries. *Journal of Food Security, 2*(1), 33–41.

Singh, J. (2009). Labor Law and Special Economic Zones in India, CSLG/WP/08.

Singh, K. (2013). Overview of Special Economic Zones (SEZs) with a Special Reference to Haryana. *Global Journal of Management and Business Studies, 3*(1), 1235–1240.

UN Department of Economic and Social Affairs Report. (2015). World Population Prospects: The 2015 Revision.

UNDP. (2015). If Africa Builds Nests, Will the Birds Come? Comparative Study on Special Economic Zones in Africa and China. Working Paper Series No. 6.

World Bank. (2008). *World Development Report on Agriculture for Development*. Washington, DC: World Bank.

Zeng, D. Z. (2015). Global Experiences with Special Economic Zones: Focus on China and Africa. Retrieved July 10, 2016, from https://openknowledge.worldbank.org/bitstream/handle/10986/21854/WPS7240.pdf?sequence=2.

http://www.sezindia.nic.in/about-ep.asp. Retrieved 13 July 2016.

http://www.worldbank.org/en/region/afr/brief/fact-sheet-doing-business-2016-in-sub-saharan-africa. Retrieved 14 July 2016.

http://www.macauhub.com.mo/en/2015/08/25/mozambique-aims-to-establish-special-economic-zones-for-agriculture. Retrieved 14 July 2016.

http://business.mapsofindia.com/sez/advantages-units-india. Retrieved 14 July 2016.

http://www.afdb.org/fileadmin/uploads/afdb/Documents/Generic.Documents/3.%20Zimbabwe%20Report_Chapter%201.pdf. Retrieved 27 July 2016.

12

Global Value Chains and Upgrading in Economic Community of West African States Countries

Anani Nourredine Mensah and Abdul-Fahd Fofana

12.1 Introduction

In an increasingly interconnected global economy, where over 70% of trade is in intermediate goods and services, integration into global value chains (GVCs) today will determine future trade and FDI patterns, as well as growth opportunities.[1]

Indeed, since the 1990s, global trade has undergone drastic changes. The falling transport and communication costs, coupled with technological advances and trade liberalization, have profoundly transformed the way goods and services are produced.

A. N. Mensah (✉)
University of Lome, Lome, Togo

A.-F. Fofana
Center for Research and Training in Economics and Management, Lome, Togo

A. Shimeles et al. (eds.), *Building a Resilient and Sustainable Agriculture in Sub-Saharan Africa*, https://doi.org/10.1007/978-3-319-76222-7_12

As a result, competition has increased and firms have been forced to review their organizational pattern and method of production (Porter 1986; Lorenzi 2005). For the most part, firms have expanded geographically in a given form (offshoring, outsourcing, etc.) in an effort to capture growth opportunities and competitive benefits; hence the emergence of what is known as GVCs. GVCs describe a decentralized and interconnected process, covering activities from the conception and design stages to manufacturing, marketing and commercialization of goods and services (Gereffi and Fernandez-Stark 2011).

This principle of fragmentation of production processes is the culmination of previous contributions relating to specialization and the international division of labour. It draws its inspiration from both the theory of international trade (Smithian and Ricardian theories and the Heckscher-Ohlin-Samuelson models, known as HOS for short) and the industrial economy approach (Porter and Competitive Advantage 1985).

This new configuration of world trade offers fresh opportunities and possibilities for structural change in developing countries, which are no longer forced to set up entire production units (Baldwin 2012; Escaith 2014) but can now fit themselves in as links in a GVC, depending on their comparative advantages, while benefiting from transfers of foreign skills and know-how (Hausmann et al. 2014).

As new avenues of economic growth, GVCs are certainly opening up new opportunities but are not by any means a panacea. For a firm to actually reap the benefits of participating in a GVC, its participation must go in tandem with the upgrading of its activities.[2] GVCs have been well documented, and their effects are the subject of many recent empirical studies. Nevertheless, few studies have been carried out on upgrading.

Humphrey (2004) conducted an analysis of upgrading in the agricultural and manufacturing sectors of a sample of developing countries. It revealed that participation in a GVC positively affects the technological capacity and the upgrading of economies. Rodrik (2006) tested the same assumption in China using a methodology based on sophistication measurement and found that participation in the GVC contributed significantly to the sophistication of Chinese exports. In the case of India, Felipe Jesus et al. (2012) also analysed upgrading in the GVC through export sophistication and diversification. They clearly showed that India's exports are well diversified and sophisticated. Bernhardt and Milberg (2011) also analysed upgrading in certain sectors (horticulture, clothing, mobile phones and tourism) of the GVC. The results highlight the existence of upgrading, with the exception of the clothing sector.

This issue has been well documented in the case of African countries. However, worth mentioning is Hidalgo (2011) who analysed upgrading in East African countries. Using the concept of product space to analyse export diversification and sophistication, he found that these countries, with the exception of Kenya, generally have poorly diversified and unsophisticated exports (all of which are located on the periphery of the product space). By measuring export sophistication, Hausmann et al. (2014) showed that exports from Uganda are poorly diversified and unsophisticated. Similar findings were made by Abdon and Felipe (2011) and Hausmann and Jasmina (2015) respectively in sub-Saharan Africa and Rwanda.

The objective of this chapter is to analyse the upgrading of African countries (especially those of Economic Community of West African States [ECOWAS]) in the GVC. This choice was motivated by two main reasons. First, West Africa is one of the most open regions in the world. However, it must be said that the region's share of international trade remains below its potential and represents 0.7% in value of world exports, compared with 0.5% of imports.[3] Moreover, in terms of upgrading, these countries have lagged behind other regional groups, which seems to suggest that the openness has contributed little to improving economic performance. Hence, the question of whether or not the position of its States in trade allows them to benefit from their integration in the world economy. Second, in 2014, these countries concluded negotiations on the Economic Partnership Agreement (EPA) with the European Union, which have led to the promotion of integration into the GVC. This justifies the choice of this zone where there are few empirical studies on the GVC theme.

This chapter is divided into three sections: (1) the first section defines GVC and upgrading concepts; (2) the second section analyses the level of participation of ECOWAS countries in the GVC; and (3) the last section analyses the upgrading of countries of the community.

12.2 Definition of the GVC and Upgrading Concepts

12.2.1 GVC Concept

In recent years, there has been a shift from trade that helps to "sell" goods to trade that helps to "make" goods (Baldwin et al. 2014). This phenomenon, formalized by the expression "global value chains", may also be comprehensible under the terms "global supply chains", "international production networks", "vertical specialisation", "outsourcing" and "production fragmentation".

The GVC concept is promoted by Porter (1986) who describes it as a set of interdependent and coordinated activities allowing the creation of identifiable and measurable value if possible. The value chain encompasses all backward and forward activities leading to the production of a product or service (Porter 1986). A GVC refers to when these activities are fragmented across sites and borders (Lunati 2008).

It also refers to the full range of activities which are required to bring a product or service from design through the various phases of production and delivery to final consumers and final disposal after use (Kaplinsky 2004).

GVCs refer to the interconnected production process that goods and services undergo from conception and design through production, marketing and distribution (Gereffi and Fernandez-Stark 2011).

In this research, we have adopted the simple notion proposed by Lunati (2008) and which seems to capture the meaning of most of the above definitions. According to Lunati, GVCs are international supply chains characterized by fragmentation of production activities across sites and borders.

12.2.2 Upgrading Concept

A company is upgraded in the GVC to which it already belongs when it creates more value added (Gereffi et al. 2001). In a value chain, various types of upgrading may be distinguished (Humphrey and Schmitz 2000):

"process upgrading", "product upgrading", "functional upgrading" and "chain upgrading".

- "Process upgrading" takes place when there is an improvement in the production process, allowing more efficient transformation of inputs into outputs. The company is therefore able to perform tasks in a more efficient way and with a lower imperfection rate than its competitors.
- "Product upgrading" takes place when the company can introduce new products, modify the design, improve the quality and supply an end product that has a higher value added by virtue of its higher level of sophistication.
- "Functional upgrading" occurs when other stages of production in the GVC can be accessed. In this case, the company is able to offer competitive products with greater value added. This means that changes are made upstream and downstream of the production process.
- "Chain upgrading" or "inter-chain upgrading" corresponds to movement from one industry to another. It thus occurs when a company is able to refocus or position its activities in new GVCs with higher value added. Very often, greater integration into the GVC is also referred to as "institutional upgrading".

A company can then upgrade in the GVC either by optimizing the value of its supply, developing a strategy for adding services to its range of products, or by implementing a customer strategy through stronger relationships with its clientele (Lahille et al. 1995).

12.3 GVC Participation

12.3.1 GVC Participation Measurement

To measure a country's participation in the GVC, it is necessary to know the sources and destinations of the value added of the products. Two indicators are usually used to measure a country's GVC participation: the "*backward integration*" index and the "*forward integration*" index.

Backward integration measures the share of inputs imported by a country and used in local production for export purposes, or the share of foreign value added (FVA) incorporated in a country's exports. Forward integration measures the share of domestic value added (DVA) in exports from other countries. The GVC participation index is the sum of these two indicators expressed as a percentage of gross exports (Koopman 2011).

12.3.2 Level of Participation of ECOWAS Countries in the GVC

Africa accounts for a modest but growing share of value-added trade (from 1.4% in 1995 to 2.2% in 2011).[4] West Africa is the third-best region in Africa in terms of GVC integration, but the integration is strongly driven by forward integration. With just under USD 40 million in 2011, West Africa accounts for about 15% of Africa's GVC participation, with only a quarter being backward integration (Fig. 12.1).

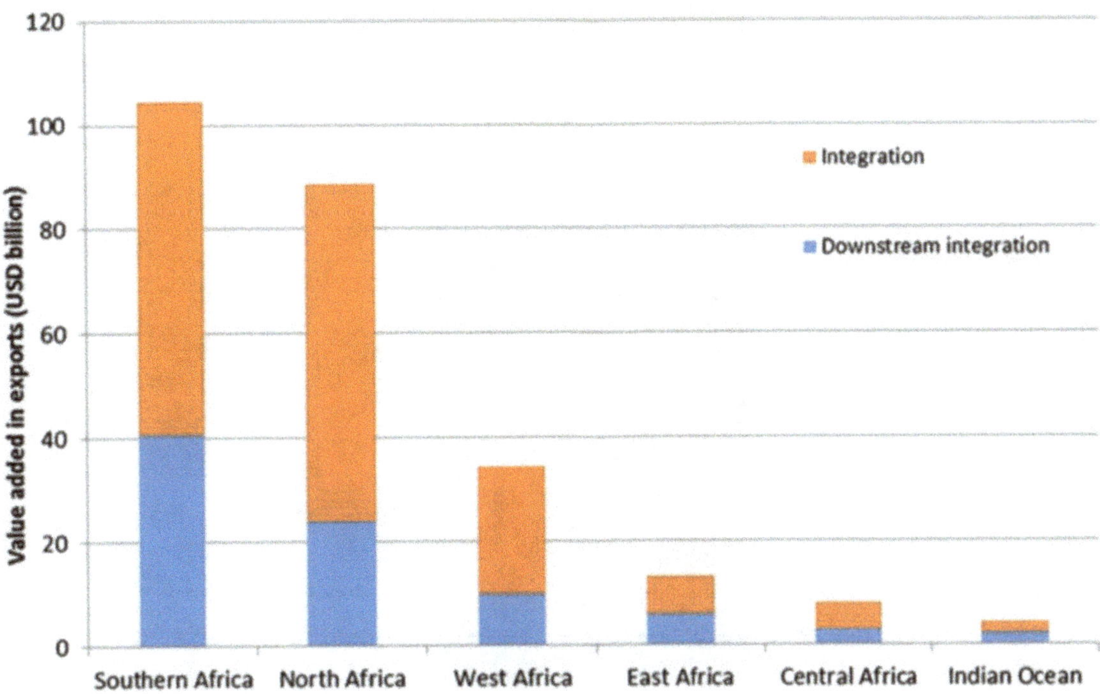

Fig. 12.1 Integration of African regions in GVCs, 2011. Source: Authors' elaboration based on AfDB et al. (2014) (from UNCTAD-EORA-GVC data)

Europe and Asia are the continents with which ECOWAS trade most in terms of value added. West African inputs in the GVC (Fig. 12.2) are mainly destined for Europe and Asia, which respectively absorb 60% and 12% of West African products integrated downstream of the value chain. Regarding backward integration, Europe is also West Africa's leading supplier (Fig. 12.3), with a share of around 40%. Asia comes second with a share of about 32%.

Figures 12.4 and 12.5 illustrate the FVA incorporated in the exports of ECOWAS countries and the export value added (EVA) of these countries for 1990, 1995, 2000, 2005 and 2012, respectively. For most of these countries, the levels of FVA and the EVA are very low. However, Nigeria and countries such as Côte d'Ivoire, Ghana and, to a lesser extent, Senegal have an acceptable level of trade value added. For these countries, both foreign and domestic trade value added increased over the 1990–2012

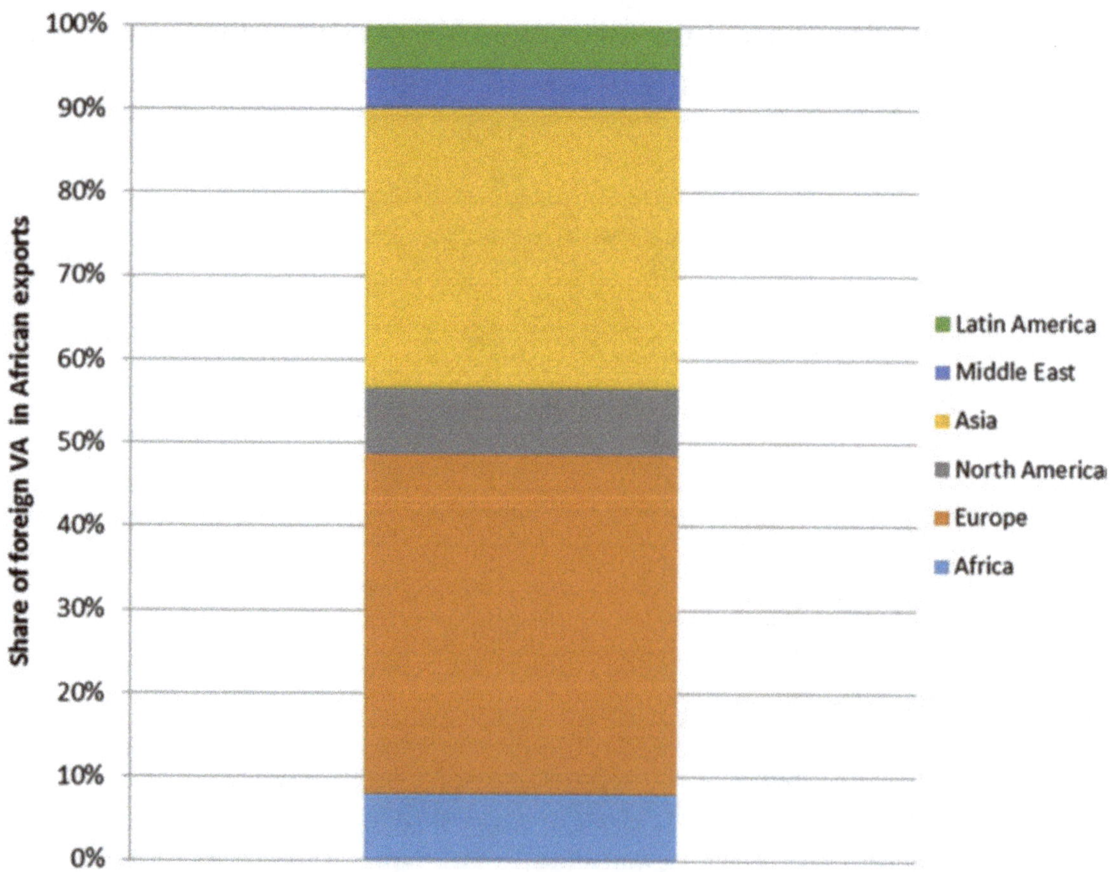

Fig. 12.2 Sources of intermediary products in West Africa. Source: Authors' elaboration based on AfDB et al. (2014) (based on UNCTAD-EORA-GVC data)

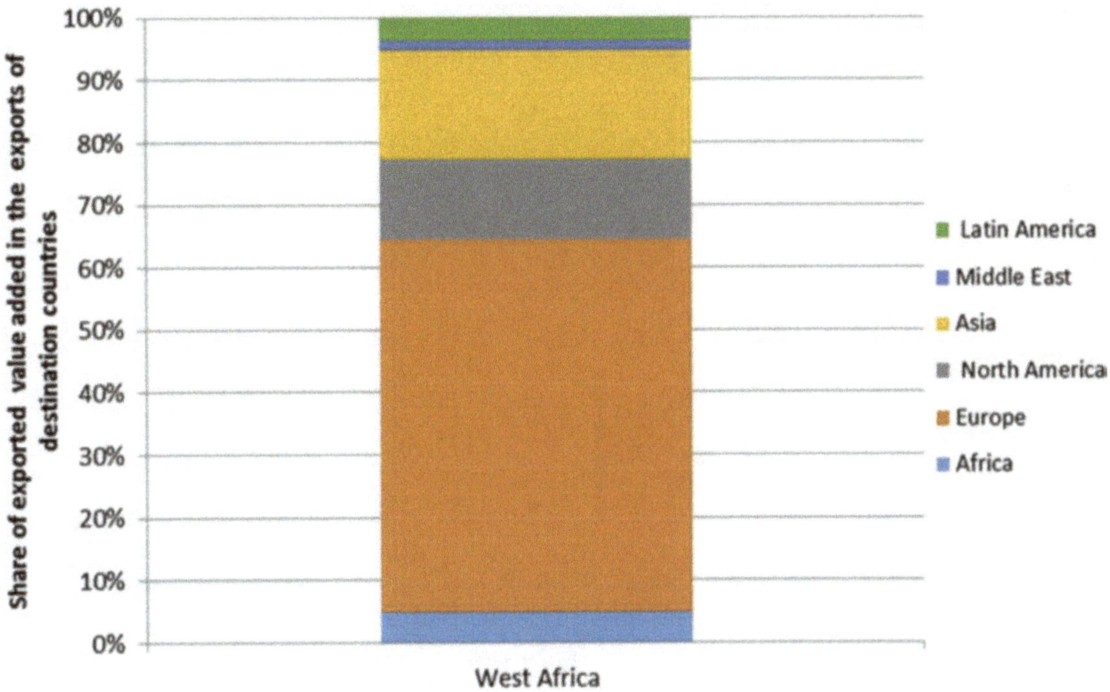

Fig. 12.3 Destinations of intermediary products in West Africa. Source: Authors' elaboration based on AfDB et al. (2014) (based on UNCTAD-EORA-GVC data)

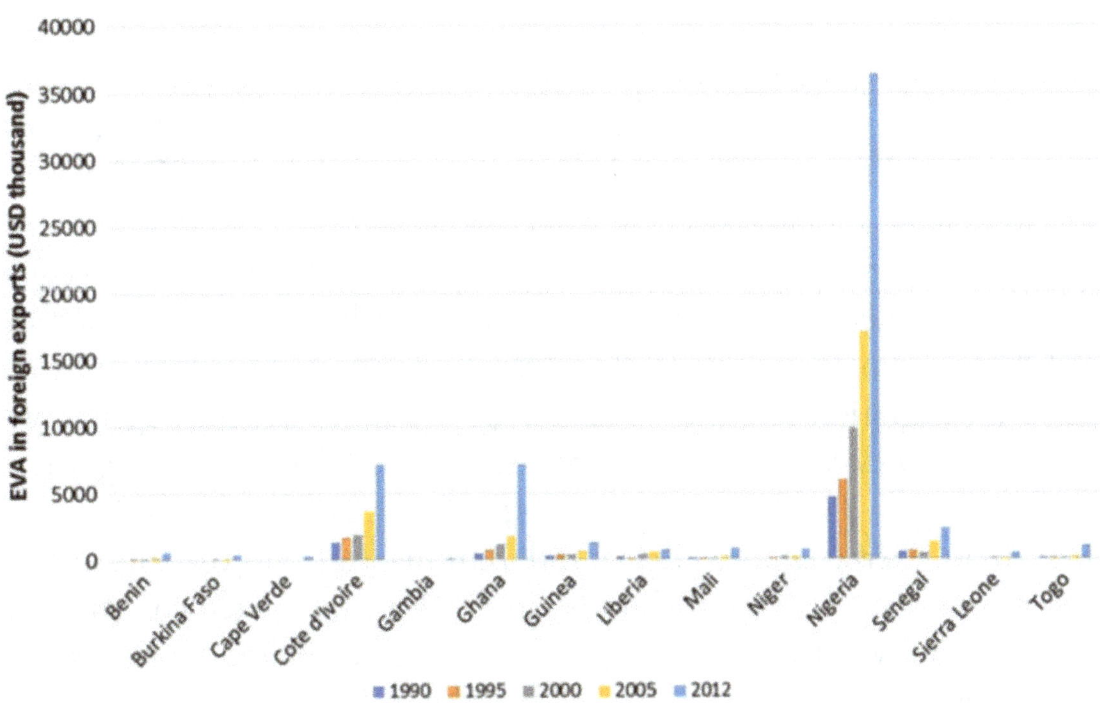

Fig. 12.4 EVA content of foreign exports (USD thousand). Souce: Authors' elaboration based on UNCTAD-EORA-GVC data

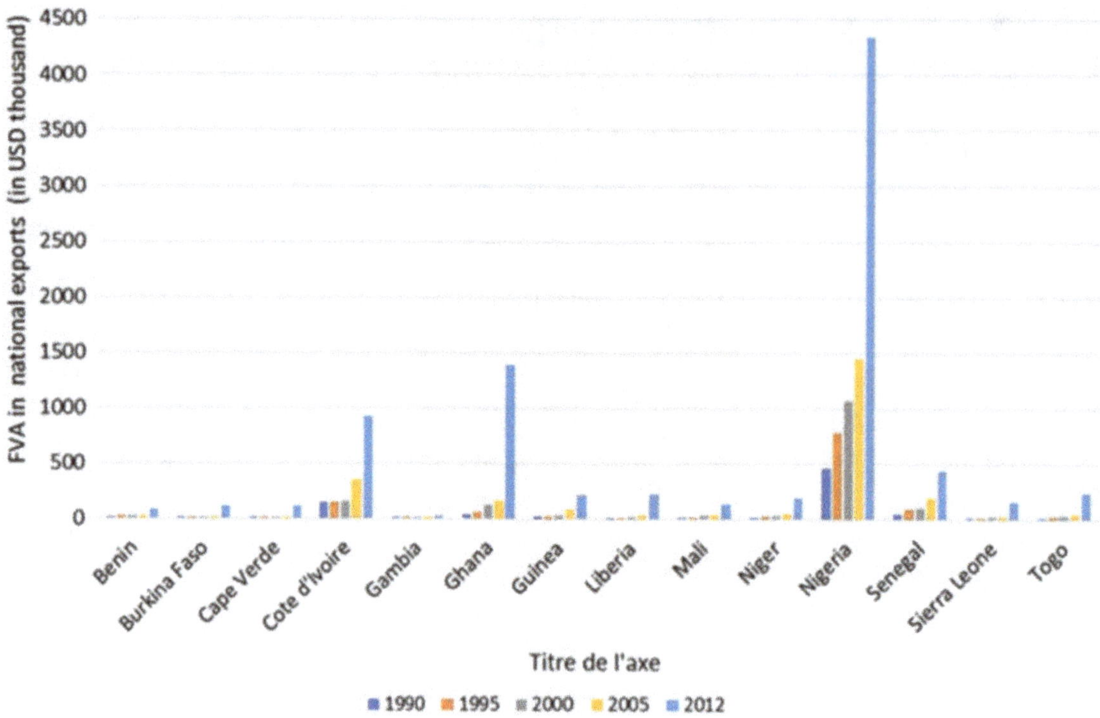

Fig. 12.5 FVA content of national exports (USD thousand). Source: Authors' elaboration based on UNCTAD-EORA-GVC data

period (for Nigeria, for example, the exported value added rose from about USD 5 million in 1990 to about USD 35 million in 2012, compared with the foreign value added which increased from USD 0.5 million to USD 4.5 million over the same period).

The average participation of West Africa in the GVC conceals disparities between member countries. Taken individually, the participation of ECOWAS countries in the GVC is very low, driven by a high level of forward integration. Guinea, Ghana and, to a lesser extent, Nigeria, are the most integrated countries downstream of the GVC, with integration levels of 41%, 32% and 30%, respectively. In terms of backward integration, Togo, Sierra Leone, Ghana and Burkina Faso are the most integrated countries. Benin and Gambia are the least integrated countries in the community with a total integration level of 27% and 29%, respectively (Fig. 12.6).

In short, this analysis shows that ECOWAS countries effectively participate in the GVC, but the participation is strongly driven by primary commodity exports, which may limit any possibility of upgrading in the GVC.

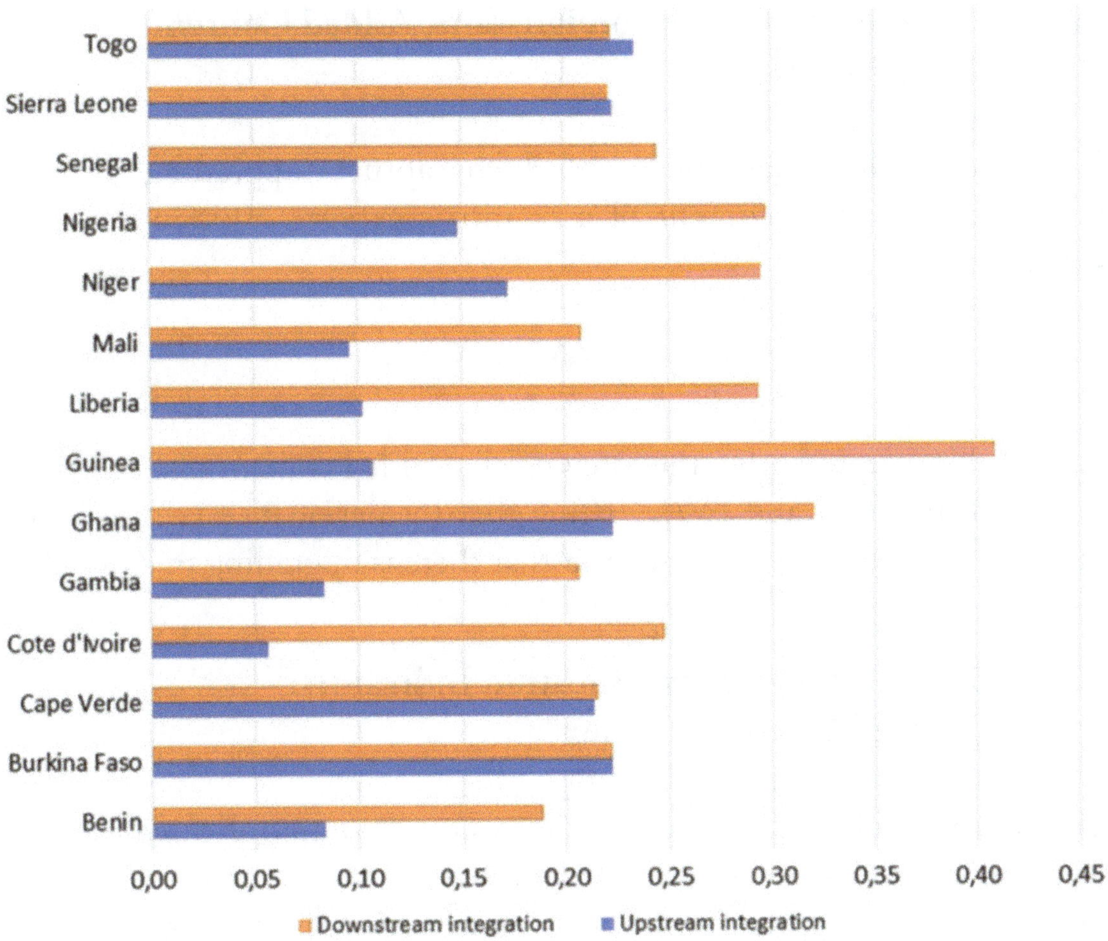

Fig. 12.6 Integration of ECOWAS countries in the GVC, 2011 (%). Source: Authors' elaboration based on AfDB et al. (2014) (based on UNCTAD-EORA-GVC data)

12.4 Upgrading of ECOWAS Countries

12.4.1 Mythological Data

To analyse the different aspects of upgrading, we have used a number of indicators. We began by capturing the upgrading through the change in FVA content of a country's exports. Then we have adopted the approach used by Cottet et al. (2012) to capture the upgrading of diversification, sophistication and export base renewal indicators.

12.4.1.1 Indicator of the Foreign Value Added Content of Exports

Here, we developed an approach which captures upgrading through increase in the FVA content of a country's exports. This indicator is given by:

$$I_{\text{FVA}} = \text{FVA}_t - \text{FVA}_{t-1} \quad (12.1)$$

With FVA_t representing the foreign value added content of a country's exports at period t, and FVA_{t-1} being that for period $t - 1$. When this indicator is positive, we may suspect that there is upgrading in the GVC.

12.4.1.2 Traditional Diversification Indicators: The Hirschman Index

The Hirschman index is one of the indices most commonly used to measure the weight of each sector in total exports (Cadot et al. 2013). According to this approach, the less a country depends on a limited number of export goods, the more it is considered diversified. Conversely, when a product accounts for a huge portion of a country's exports, the country is considered concentrated. This index is calculated as follows:

$$H_1 = \sqrt{\sum_{i=1}^{N} \left(\frac{x_i}{X} \right)^2}, \quad (12.2)$$

where (x_i) denotes the value of the exports the good i, X the total exports $(X = \sum_{i=1}^{N} x_i)$ and N the number of product groups. It is then standardized for easier reading:

$$NH_1 = \frac{H_1 - \sqrt{\frac{1}{N}}}{1 - \sqrt{\frac{1}{N}}} \tag{12.3}$$

The NH value closest to 1 represents the highest concentration/lowest diversification and vice versa. When this indicator is equal to 1, the country is entirely dependent on a single export product.

12.4.1.3 Export Sophistication Measurement Indicator

The capacity to incorporate technological content into exports is not limited to increasing the degree of diversification. This leads us to a new indicator for measuring the degree of exports sophistication. This indicator assesses the level of industrial exports as a share of the population (Cottet et al. 2012) and is calculated as follows:

$$I_{\text{inndus}} = \frac{\sum_{i \in K} x_i}{\text{POP}}, \tag{12.4}$$

where K denotes the subgroup of industrial products and POP,[5] the country's population. This indicator isolates export products other than primary products (agricultural or extractive), which make up the bulk of a country's exports.

12.4.1.4 Capacity to Export New Products: Extensive Margin and Intensive Margin

A lot of publications break down export growth according to the appearance of new export lines (extensive margin) or according to the increase in the export of already existing products (intensive margin) (Melitz 2003). Depending on which margin dominates the other, export growth can stem from either diversification or specialization. Indeed, when the extensive margin dominates the intensive margin, the upgrading of products results in exports diversification (Cadot et al. 2011). On the other hand, when the intensive margin accounts for most of the exports, this may reflect specialization in the export base (Helpman et al. 2008).

However, the launch of new export products is not necessarily an end in itself, nor sufficient to ensure export diversification. The new products launched must therefore consolidate over time. We thus witness an alternation of diversification and concentration phases, causing the so-called *Big Hits* phenomena whereby export growth is driven by a few flagship products (Easterly and Reshef 2010).

The latter indicator, inspired by the works of Easterly and Reshef (2010) and Amiti and Freund (2010), breaks down export growth as follows:

$$\frac{x_t - x_{t-1}}{x_{t-1}} = \frac{t_t - t_{t-1}}{x_{t-1}} + \frac{n_t - d_{t-1}}{x_{t-1}} \quad (12.5)$$

where $\frac{t_t - t_{t-1}}{x_{t-1}}$, measures the intensive margin and $\frac{n_t - d_{t-1}}{x_{t-1}}$ the extensive margin. The intensive margin is measured by the increase in "traditional" exports (referred to as t) exported at two periods $t-1$ and t. The extensive margin is measured by the increase in new exports, or the difference between new exports (n) at period t and products that have disappeared from the exports (d) since period $t-1$, with (x_{t-1}) denoting total exports at period $t-1$, and (x_t) the exports at period t.[6]

To differentiate export flagship products from nascent products, we break down the intensive margin according to three types of goods[7]: low-intensity export products (t_F), medium-intensity export products (t_M) and flagship export products (t_P). The first account for less than 2% of the country's total exports, while the last represent between 2% and 10% of total exports and the last more than 10% of total exports. Our equation is thus rewritten as follows:

$$\frac{x_t - x_{t-1}}{x_{t-1}} = \frac{t_{F_t} - t_{F_{t-1}}}{x_{t-1}} + \frac{t_{M_t} - t_{M_{t-1}}}{x_{t-1}} + \frac{t_{P_t} - t_{P_{t-1}}}{x_{t-1}} + \frac{n_t - d_{t-1}}{x_{t-1}} \quad (12.6)$$

This breakdown of the intensive margin makes it possible to determine whether the export growth is due to the flagship products or rather the *Big Hits* phenomena.

UNCTAD's EORA-GVC (2014) database is used to analyse the FVA indicator of a country's exports, while the agency's Commodity Trade Statistics Database (Comtrade) and World Integrated Trade Solution (WITS) database are used to calculate the other indicators. Population data come from the World Bank's World Development Indicators (WDI) database.

12.5 Results

The results are presented according to the methodological approach. Thus, we first present the results of the EVA indicator followed by the diversification, sophistication and export base renewal indicators.

12.5.1 Indicator of the Foreign Value Added Content of Exports

In most ECOWAS countries, between 1995 and 2011, GVC growth occurred in tandem with the FVA of exports. We note that most countries are in the upper right quadrant, that is to say, they have increased the share of FVA content of their exports as well as the share of local value added content of exports relative to GDP. This suggests that from 1995 to 2011, upgrading in the GVC became more pronounced in most ECOWAS countries. (Fig. 12.7). It should be noted, however, that this indicator is not sufficient to characterize a country's upgrading in GVC, given that it does not allow for export diversification analysis.

12.5.2 Traditional Indicators: Hirschman index

The reading of this index (Annex 1) shows that ECOWAS countries are characterized by low diversification levels. Generally, most of the countries have highly concentrated goods exports, as shown by the diversification

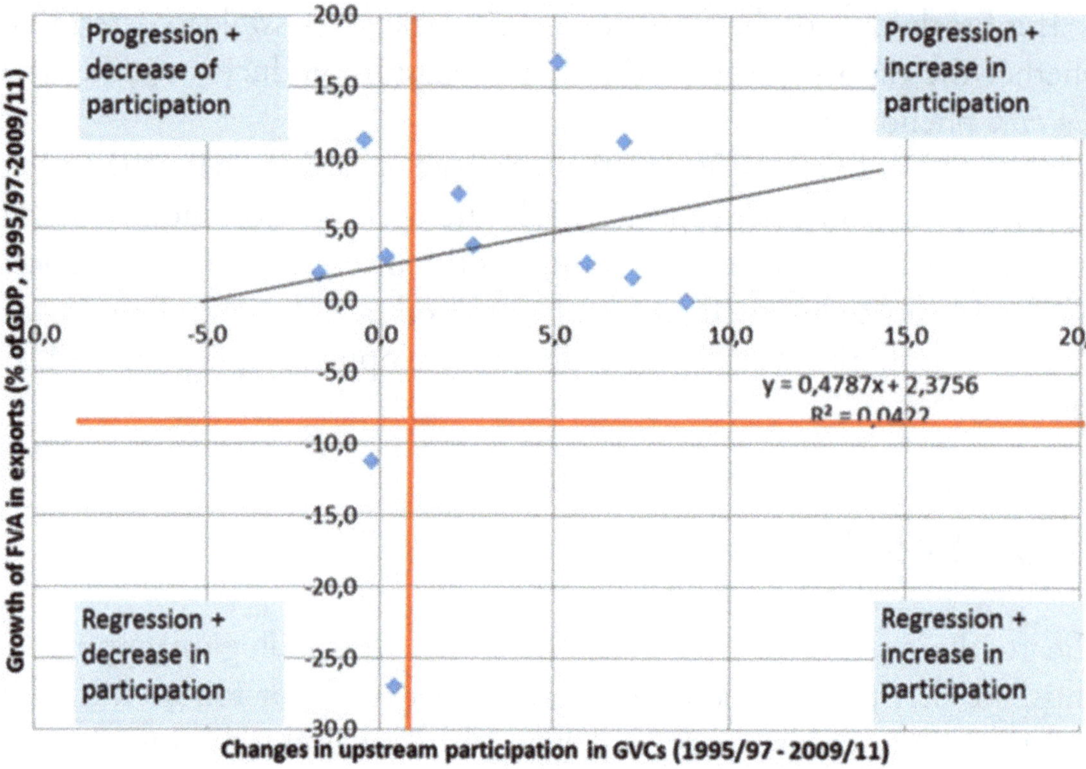

Fig. 12.7 Participation in GVC and growth of FVA in exports as a percentage of GDP, 1995/1997, compared with 2009/2011. Source: AfDB et al. (2014) (based on UNCTAD-EORA-GVC data)

index, which has an average value of more than 0.7 for all countries. These countries have therefore not been successful in upgrading by directing their conventional exports towards new, more dynamic and more promising sectors. Nigeria appears to be the most concentrated ECOWAS economy, with an average diversification index of 0.85 over the study period. This position enjoyed by Nigeria could be attributed to a strong concentration on oil export, which presumably makes up the lion's share of the country's exports.

Exports from all of these countries are highly concentrated on a limited number of low-tech products, which is confirmed by the statistics in the table provided in Annex 2. This table presents, for each country, the average share of the top five exported products, compared with total exports over the period 2010–2014. For all countries, the share of the leading export product in the total exports averages 47% and accounts for more than half of total exports in 5 of the 14[8] ECOWAS countries for which data are available. In some countries, the largest export product dominates total exports (Guinea-Bissau: 96%, Nigeria: 73%, Mali: 72%). It is not less dominant in other countries such as Burkina Faso (52%, Gold), Niger (50%, Uranium) or Gambia (50%, artificial filament fabrics). Togo, Senegal and Côte d'Ivoire are exceptions, with the share of their dominant export products accounting for less than 50% of total exports (or 16%, 16% and 23% respectively).

In Benin, cotton, refined petroleum, cashew nut, rice and gold are the main export commodities over the period under review and account for about 54% of total exports. Gold, cotton, sesame seeds, cashew nuts and oilseeds are Burkina Faso's main export products and account for about 76% of total exports. The growth of mining activities in Burkina Faso, especially gold exports, is robust. Invigorated by new discoveries of deposits and a generous tax system designed to attract foreign investors, gold production represents 52% of the country's total exports between 2010 and 2014. In Côte d'Ivoire, a review of the diversification index (Annex 1) shows that the economy is also highly concentrated. Cocoa, oil and rubber were, on average, the most exported conventional products over the period 2010–2014. In Cape Verde, tuna is the leading export product, accounting for 34% of total exports. The five products exported during the period under review represented 84% of total exports, on average. Artificial filament fabric exports make up about 50% of total Gambian exports (the top five export products account on average for 61% of total exports). In Guinea-Bissau, the pattern of goods exports makes them highly concentrated on a single product. Raw cashew nuts remain the main export product, with an average share of about 96%. Aluminium, gold, postage stamps and rubber account for 90% of Guinea's exports. Gold and cotton are the main exports of Mali and together account for 78% of total exports.

Uranium is Niger's traditional export product (50%), with the top five exports estimated to make up 79% of the country's total exports. Oil represented, on average, 81% of Nigeria's exports between 2010 and 2014. In Sierra Leone, the range of exports consists mainly of tin, ethylene-vinyl acetate copolymer, packaging and cocoa husks, representing 87% of total exports. Oil, gold, phosphoric acid, cement and fresh fish make up 48% of Senegal's total exports. Togo is rich in mineral resources, which places the country at the forefront of economic diversification within the community. Cotton, cement and phosphate production was estimated at 39% of total exports over the period 2010–2014 (see Annex 2). In summary, ECOWAS economies are characterized by an export pattern that is highly concentrated on natural resources, which confirms the results of the diversification index.

12.5.3 Export Sophistication Measurement Indicator

Table 12.1 shows that the level of sophistication is relatively low in ECOWAS countries, that is, their industrial exports level is low relative to the population size (compared with South Africa). Nevertheless, these countries do not form a homogeneous block of industrial product exporters. There are significant differences between these countries, which may be divided into two groups: (1) countries with the highest industrial export values relative to their populations (Ghana, Côte d'Ivoire, Nigeria, Cape Verde, Senegal, Niger, Togo and Sierra Leone) and (2) countries that export very insignificant or no industrial products (Guinea-Bissau, Gambia, Burkina Faso, Benin and Mali). These results confirm the low level of upgrading (functional upgrading and process upgrading) in ECOWAS countries, as their exports have very little technological value added. That is partly due to the fact that these countries have a low level of skilled labour, which limits all possibilities of technology transfer.

Table 12.1 Average level of industrial exports relative to population (in USD per capita) between 2010 and 2014

Country	Industrial exports/ population	Country	Industrial exports/ population
Benin	4.93	Mali	6.17
Burkina Faso	2.06	Niger	29.63
Cape Verde	47.65	Nigeria	91.53
Côte d'Ivoire	222.12[a]	Senegal	45.42
Gambia	1.39	Sierra Leone	15.12
Ghana	272.57[a]	Togo	27.75
Guinea-Bissau	0.10	South Africa	377.12

Source: UNCTAD and IMF, calculated by the authors
[a]Relatively more sophisticated

The sophistication indicator can, however, be supplemented by analysis of the sub-sectors to which these products belong. The question is whether these products belong to new export lines or to existing ones.

12.5.4 Capacity to Export New Products: Extensive Margin and Intensive Margin

The low trade diversification level does not necessarily mean that the exported products are stable. One may witness the creation of new products, an upturn in the sector or a recomposition of the export base. Product upgrading in ECOWAS countries between the period 2000–2002 and 2012–2014 is analysed by breaking down the growth of exports into an intensive margin (increase in exports of traditional products) and an extensive margin (net creation of new export products).[9] The results (Table 12.2) show a strong heterogeneity between countries. For some countries, the exports tripled (Côte d'Ivoire, Cape Verde, Guinea, Mali, Senegal and Togo), for others, they increased fivefold (Niger, Gambia, Nigeria) and even tenfold (Burkina Faso) and for others still, they doubled (Benin).

The exports from ECOWAS countries[10] have experienced an average growth rate of about 350%. The average extensive margin for these countries (228%) is nearly twice as high as the intensive margin (122%). The increase in exports is due, on average, more to the net creation of new exports than to an increase in traditional exports. However, this result changes once one begins reasoning in terms of the median levels: the median growth rate of these countries is about 250%, with an extensive median margin of 114% and an intensive median margin of 131%. In this case, the growth of exports would then be more dependent on the traditional products, which, therefore, means that there is no product upgrading.

A more detailed analysis of the results of the intensive margin shows that for most countries of the community (Benin, Burkina Faso, Côte d'Ivoire, Guinea, Mali, Nigeria, Senegal and Togo), growth is driven by moderately and intensely exported products. The main products (column g) are the most buoyant, except for Togo, whose leading export (phosphate)

Table 12.2 Product upgrading: decomposition of export growth between 2000–2002 and 2012–2014 by type of margin in %

	Extensive margin				Intensive margin			Total export growth
	Net creation	New	Disappeared	Total	Low	Average	Flagship	
	a = b – c	b	c	d = e + f + g	e	g	g	h = a + d
Gambia	588	645	57	–14	0	29	–43	574
Burkina Faso	1012	1023	11	279	0	77	202	1291
Cape Verde	243	316	73	4	0	31	–27	247
Ghana	325	343	18	4	0	0	4	329
Nigeria	68	69	1	3	–0.02	0	3	71
Benin	33	43	10	58	0	26	32	91
Côte d'Ivoire	79	96	17	174	0	15	159	253
Guinea	4	5	1	201	0	38	163	205
Mali	14	33	19	203	0	20	183	217
Niger	160	200	40	287	0	0	287	447
Senegal	149	193	44	152	0	71	81	301
Togo	62	81	19	110	0	76	35	172

Source: Authors' calculation based on WITS data (6 digit classification)

experienced a decline in favour of cotton, which was not as yet included among flagship products for the period 2000–2002. The intensive margin of low export products is virtually non-existent, with the exception of Nigeria, which has witnessed a decline in these products. This shows the difficulties that ECOWAS countries have in promoting and supporting their emerging exports in the medium and long term. This result goes to confirm the one obtained by Cottet et al. (2012) for franc-zone countries. Hausmann and Rodrik (2003) have also shown that least developed countries are finding it hard to overcome the barriers hindering the launch of new export lines.

An analysis of the results by country makes it possible to identify characteristics specific to each country or group of countries:

Product upgrading in Gambia was driven only by the extensive margin (column a). Indeed, the intensive margin for flagship exports (column g) experienced a sharp decline, reflecting the process of recomposition of this country's export base. Traditional export products (peanut and peanut oil) have indeed given way to new products (fabrics and cashew nuts).

The export growth experienced by Burkina Faso, Cape Verde, Ghana and Nigeria has been largely boosted by the increase in new products (412 percentage points on average for column a), and also by the intensive margin (column d), which contributed an average of 72.5 percentage points. However, Burkina Faso's export growth is much higher (1291 percentage points) than that of other countries (329 percentage points for Ghana, 247 for Cape Verde and 71 for Nigeria). Burkina Faso has witnessed an increase in the export of cotton (202 percentage points—column g), which has been the flagship export product since the colonial era. Despite this solid traditional base, Burkina Faso has evidently found a new export line brought about by the very rapid development of gold mining activity, reflecting an upturn in the sector. Gold production rose from a negligible volume in 2007 to almost 39 tonnes in 2013, or 71% of exports (IMF 2014)—enabling the extensive margin to contribute to the overall export growth to the tune of 1012 percentage points (column a).

The other countries (Benin, Côte d'Ivoire, Guinea, Mali, Niger, Senegal and Togo) have largely benefited from the increase in the intensive margin

(169 percentage points on average), compared with 72 percentage points on average for the extensive margin, which, in actual fact, attests to the intensification process being undergone by traditional export products. Senegal, however, stands out from the other countries by dint of the relatively homogeneous distribution of the extensive margin (149 percentage points) and the intensive margin (152 percentage points). Thus, the country has succeeded in creating new export sectors which, in the medium term, have remained in the export base (as in the case of gold exports and Portland cement—see Annex 2).

12.6 Conclusion

The analyses in this chapter show that external trade in ECOWAS is characterized by a strong expansion trend (increase in exports and imports). Sustained demand for commodities has undoubtedly stimulated the development of trade, particularly with emerging countries. Even though ECOWAS trade has risen sharply, it remains below the potential of the region when it comes to positioning in the GVC.

The trade pattern shows a dependence on commodity exports, which is a barrier to better integration in the GVC. The participation of these countries in the GVC is strongly driven by the export of primary products, which has somewhat limited the chances of upgrading in the value chain. Even though some countries have managed to create new export lines, upgrading analysis (through diversification and sophistication indicators) shows that exports from ECOWAS countries are considered to be highly concentrated on a limited number of low-tech products.

The results of this study highlight the need for effective public intervention to improve the international competitiveness of these countries and promote new products abroad. This will involve investing in infrastructure and supporting export companies. To take advantage of their integration into the world economy, we recommend more backward integration for these countries in the GVC. We also believe that integration of national productions would capture more value added through the sophistication and diversification of production.

Annex 1

Table 12.3 Diversification index of ECOWAS countries (1995, 2000, 2005, 2010 and 2014)

	1995	2000	2005	2010	2014
Benin	0.77	0.81	0.79	0.75	0.76
Burkina Faso	0.80	0.75	0.82	0.83	0.76
Cape Verde	0.61	0.66	0.71	0.72	0.70
Côte d'Ivoire	0.82	0.81	0.73	0.73	0.74
Gambia	0.79	0.76	0.70	0.75	0.76
Ghana	0.83	0.81	0.82	0.79	0.75
Guinea	0.86	0.85	0.85	0.82	0.80
Guinea Bissau	0.69	0.67	0.66	0.76	0.77
Liberia	0.77	0.83	0.85	0.71	0.82
Mali	0.76	0.81	0.82	0.84	0.84
Niger	0.77	0.85	0.78	0.79	0.83
Nigeria	0.89	0.88	0.86	0.81	0.81
Senegal	0.81	0.77	0.69	0.76	0.73
Sierra Leone	0.71	0.66	0.68	0.69	0.86
Togo	0.74	0.75	0.72	0.72	0.69

Source: UNCTAD database

Annex 2

Table 12.4 Share of the five leading exports products of ECOWAS countries in total exports (in %; 2010–2014 average)

Share of leading products at less than 50% of total exports		
Country	Main products	Share in exports (%)
Benin	Cotton	30
	Refined oil	9
	Cashew nuts	8
	Rice (Ground)	4
	Gold	3
Côte d'Ivoire	Cocoa bean	23
	Refined oil	14
	Crude oil	9
	Rubber	6
	Sawn timber	5

(continued)

Table 12.4 (continued)

Share of leading products at less than 50% of total exports		
Country	**Main products**	**Share in exports (%)**
Cape Verde	Thons	34
	Prepared or preserved mackerel	24
	Prepared or preserved fish	13
	Shoe tops, other than leather	7
	Fresh fish	5
Ghana	Transformed gold	33
	Crude oil	18
	Cocoa beans	12
	Butanes	10
	Gold	7
Guinea	Aluminium ores	45
	Gold	31
	Postage stamps, tax stamps and the like	10
	Aluminium oxide	3
	Rubber	1
Senegal	Refined oil	16
	Gold	12
	Phosphoric acid	9
	Portland cement	8
	Fresh fish	4
Togo	Cotton	16
	Cement (clinker)	9
	Portland cement	7
	Phosphates	7
	Make-up and skin care products	4
Burkina Faso	Gold	52
	Cotton (unginned)	16
	Sesame seeds	4
	Cashew nuts	2
	Oilseeds	2
Gambia	Artificial filaments fabrics	50
	Cashew nuts	4
	Clothing and other items to wear	3
	Groundnut oil	3
	Refined oil	3
Guinea Bissau	Cashew nuts	96
	Cranes	1
	Cotton	0
	Crude oil	0
	Paper pulp	0

(continued)

Table 12.4 (continued)

Share of leading products at less than 50% of total exports		
Country	**Main products**	**Share in exports (%)**
Mali	Gold	72
	Cotton (ginned)	6
	Cotton (unginned)	4
	Mineral or chemical fertilizer (with nitrogen)	3
	Mineral or chemical fertilizer (without nitrogen)	2
Niger	Uranium	50
	Crude oil	22
	Clothing and other items to wear	3
	Radioactive products	2
	Gold	2
Nigeria	Crude oil	73
	Refined oil	8
	Gas	5
	Rubber	4
	Cocoa beans	1
Sierra Leone	Tin	73
	Ethylene-vinyl acetate copolymer	11
	Articles for packaging of goods	2
	Cocoa shells	2
	Automobiles with reciprocating piston engine	1

Source: Authors' calculation based WITS data (HS classification)

Notes

1. According to the study entitled "Global Value Chains: Challenges, Opportunities and Implications for Policy".
2. A company undergoes upgrading in the GVC when it creates higher value added by moving away from low-tech activities (Gereffi et al. 2001).
3. Report of the African Centre for Trade, Integration and Development (ACACID), 2012.
4. AfDB et al. (2014).
5. The population is chosen instead of GDP because of the marked differences in the production pattern of ECOWAS countries. Oil producing countries have higher GDP per capita than others.

6. $t - 1$ covers the period 1990–1992 and t corresponds to the period 2010–2012. This makes it possible to control the exceptional exports of new products and irregularities of declaration in the calculation of the extensive margin.
7. We drew inspiration from Cottet et al. (2012).
8. Data are not available for Liberia.
9. It may be said that the country is upgrading (product upgrading) if the extensive margin is wider than the intensive margin.
10. Those for which the data needed for calculation were available: Benin, Burkina Faso, Cap Vert, Côte d'Ivoire, Gambia, Ghana, Guinea, Mali, Niger, Nigeria, Senegal and Togo.

References

Abdon, A., & Felipe, J. (2011). *The Product Space: What Does it say About the Opportunities for Growth and Structural Transformation of Sub-Saharan Africa?*

AfDB, OECD, & UNDP. (2014). *African Economic Outlook*. Paris: OECD Publishing.

Amiti, M., & Freund, C. (2010). The Anatomy of China's Export Growth. In *China's Growing Role in World Trade* (pp. 35–56). University of Chicago Press.

Baldwin, D. A. (2012). Power and International. In W. Carlsnaes, T. Risse, & B. A. Simmons (Eds.), *Handbook of International Relations* (pp. 273–297). London: Sage.

Baldwin, J. R., Yan, B., & others. (2014). *Global Value Chains and the Productivity of Canadian Manufacturing Firms*. Statistics Canada.

Bernhardt, T., & Milberg, W. (2011). *Economic and Social Upgrading in Global Value Chains: Analysis of Horticulture, Apparel, Tourism and Mobile Telephones*.

Cadot, O., Carrère, C., & Strauss-Kahn, V. (2011). Export Diversification: What's Behind the Hump? *Review of Economics and Statistics, 93*(2), 590–605.

Cadot, O., Carrere, C., & Strauss-Kahn, V. (2013). Trade Diversification, Income, and Growth: What Do We Know? *Journal of Economic Surveys, 27*(4), 790–812.

Cottet, C., Madariaga, N., & Gou, N. J. (2012). La Diversification Des Exportations En Zone Franc: Degré, Sophistication et Dynamique. *AFD, Macroéconomie et Développement, 3*.

Easterly, W., & Reshef, A. (2010). *African Export Successes: Surprises, Stylized Facts, and Explanations*. National Bureau of Economic Research.

Escaith, H. (2014). Mapping Global Value Chains and Measuring Trade in Tasks. In B. Ferrarini & D. Hummels (Eds.), *Asia and Global Production Networks: Implications for Trade, Incomes and Economic Vulnerability*. Cheltenham and Metro Manila: Asian Development Bank and Edwar Elgar Publishing.

Felipe, J., Kumar, U., Usui, N., & Abdon, A. (2012). Why Has China Succeeded? And Why it Will Continue to Do So. *Cambridge Journal of Economics, 37*(4), 791–818.

Gereffi, G., & Fernandez-Stark, K. (2011). *Global Value Chain Analysis: A Primer*. Center on Globalization, Governance & Competitiveness (CGGC), Duke University, North Carolina, USA.

Gereffi, G., Humphrey, J., & Kaplinsky, R. (2001). Introduction: Globalisation, Value Chains and Development. *IDS Bulletin, 32*(3), 1–8.

Hausmann R., & Jasmina, C. (2015). Moving to the Adjacent Possible: Discovering Paths for Export Diversification in Rwanda. *CID Working Paper* No. 294, April.

Hausmann, R., & Rodrik, D. (2003). Economic Development as Self-Discovery. *Journal of Development Economics* 72 (2): 603–633. Rapport du FMI No. 14/215.

Hausmann, R., Cunningham, B., Matovu, J. M., Osire, R., & Wyett, K. (2014). *How Should Uganda Grow?*

Helpman, E., Melitz, M., & Rubinstein, Y. (2008). Estimating Trade Flows: Trading Partners and Trading Volumes. *The Quarterly Journal of Economics, 123*(2), 441–487.

Hidalgo, C. A. (2011). Discovering Southern and East Africa's Industrial Opportunities. Economic Policy Paper Series. *The German Marshall Fund of the United States.*

Humphrey, J. (2004). Upgrading in Global Value Chains. *International Labour Office Working Paper.*

Humphrey, J., & Schmitz, H. (2000). *Governance and Upgrading: Linking Industrial Cluster and Global Value Chain Research* (Vol. 120). Brighton: Institute of Development Studies.

Kaplinsky, R. (2004). Spreading the Gains from Globalization: What Can Be Learnt from Value-Chain Analysis. *Problems of Economic Transition, 47*(2), 74–115.

Koopman, R. (2011). *Powers1W., Wang, Z., Wei, SJ Give Credit to Where Credit Ls Due: Tracing Value Added in Global Production Chains [Z]*. NBER Working Papers.

Lahille, É., Plichon, C., Vadcar, C., & Weber, B. (1995). Les Principales Causes Des Délocalisations. In *Au-Delà Des Délocalisations. Globalisation et Internationalisation Des Firmes, Chambre de Commerce et d'industrie de Paris, Collection «Entreprise et Perspectives Économiques»*. Economica.

Lorenzi, J. H. (2005). *Mondialisation et nouvelle stratégie d'entreprise*. Paris: Paris Dauphine University.

Lunati, M. (2008). Enhancing the Role of SMEs in Global Value Chains'. *Staying Competitive in the Global Economy, 65*.

Melitz, M. J. (2003). The Impact of Trade on Intra-Industry Reallocations and Aggregate Industry Productivity. *Econometrica, 71*(6), 1695–1725.

Porter, M. E. (1986). Changing Patterns of International Competition. *California Management Review, 28*(2), 9–40.

Porter, M. E., & Competitive Advantage. (1985). *Creating and Sustaining Superior Performance*. New York: Free Press.

Rodrik, D. (2006). What's so Special About China's Exports? *China & World Economy, 14*(5), 1–19.

13

Conclusions: Enhancing the Resilience and Sustainability of the Agriculture Sector in sub-Saharan Africa

Abebe Shimeles, Audrey Verdier-Chouchane and Amadou Boly

13.1 Introduction

The decline in oil and metal commodity prices which started mid-2014 has served as an incentive for African countries to focus on agricultural issues and to make strategic choices for transforming the agriculture sector and reducing dependency on food imports (AfDB et al. 2017). In this volume on *Building a Resilient and Sustainable Agriculture in sub-Saharan Africa*, authors have reaffirmed the importance of increasing agricultural productivity, addressing the climate change challenges and promoting agro-industrialization to achieve the objectives of tackling food insecurity and industrializing Africa. In turn, this will create jobs, economic development and allow the improvement of the quality of life. Increasing agricultural productivity could be achieved through many ways, including implementation of modern technologies, appropriate land tenure and better access to land, improved agricultural mechanization and use of irrigation as well as the adoption of high-yielding crop

A. Shimeles • A. Verdier-Chouchane (✉) • A. Boly
African Development Bank, Abidjan, Côte d'Ivoire

A. Shimeles et al. (eds.), *Building a Resilient and Sustainable Agriculture in Sub-Saharan Africa*, https://doi.org/10.1007/978-3-319-76222-7_13

varieties. To further propel agricultural transformation in sub-Saharan Africa and face the climate change challenges, agriculture insurance, sustainable resource management plans and integrated rural development strategies could be used as efficient measures (AfDB et al. 2017). For agro-industrialization, emphasis should be given on increasing competitiveness through closing the infrastructure gap, skills gap, reforming regulations and institutions, deepening value chains, attracting foreign direct investment through preferential taxes and creation of industrial clusters and Special Economic Zones (SEZs).

In this volume, authors have offered valued policy recommendations aimed at enhancing resilience and sustainability of the agriculture sector. The list of policy recommendations is not exhaustive, but they have been divided into five main sections. The first set of recommendations relates to the acceleration of agricultural productivity through innovation and training. The second and the third sets of recommendations respectively deal with the improvement of policies and institutions and the adoption of innovative financing for agriculture. The fourth set of recommendations considers the strengthening of agricultural value chains at a regional level and the last one, the development of infrastructure.

13.2 Accelerating Agricultural Productivity Through Innovation and Training

First of all, as demonstrated by the green revolution and the significant acceleration of agricultural productivity in Asia,[1] farmers should be procured with agriculture inputs such as fertilizer, seeds, pesticides and equipment at reasonable cost. In addition to constrain productivity, the non-modernization of the sector makes it unattractive to youth. Despite the potential for "agri-preneural" activity, Africa's youth are often moving away from agriculture to get jobs in the informal service sector that provide few more opportunities for advancement (AfDB et al. 2017).

For Christelle Tchamou Meughoyi (Chap. 2), innovation in agriculture improves agricultural performances and increases the productivity of family farms. It not only provides benefits, it also brings about social change. Carren Pindiriri (Chap. 3) highlights the importance of adopting modern technology in agriculture to cope with climate change in sub-Saharan Africa. Drought-tolerant and water-efficient crop varieties and technologies will be fundamental requirements for developing and sustaining Africa's agriculture (Kanu et al. 2014).

Also, farmers should take advantage of recent development in Information and communication technologies (ICTs) as mobile and other information technologies help innovations in agriculture. Verdier-Chouchane and Karagueuzian (2016) review the successful new ICT-based services in the agriculture sector in Africa. Both at the pre-cultivation and post-harvest stages, ICTs such as mobile phone, geographical information system and remote sensing can be used for land registration, crop inventories, common information system platform, dissemination of information on market prices, traceability information, green practices and so on. By leveraging ICTs, Africa's agriculture can improve along with these new technologies and eventually reach green and inclusive growth faster than other developing regions.

For Adedoyin Mistura Rufai, Kabir Kayode Salman and Mutiat Bukola Salawu (Chap. 4), increasing productivity through better access, availability and efficient use of agricultural inputs by farmers also contributes to reducing gender productivity differentials. However, this should be complemented with a training program to build the capacity of farmers and enhance their resource use skills and production efficiency. In the same vein, Carren Pindiriri (Chap. 3) recommends the reduction of information asymmetry among farmers and increased publicity on modern technology through various media (radio, TV) to enhance sustainable development and poverty elimination. Education and formal training of smallholder farmers will increase their propensity to modernize their production systems.

Kanu et al. (2014) reaffirm that agricultural transformation in sub-Saharan Africa requires the strengthening of technical, financial and business management skills and capacities of the rural population. As a result of low agricultural productivity, farmers survive on subsistence income while agricultural and non-agricultural productivity gap is due to differences in skills and abilities. In essence, a movement of workers from agriculture to non-agriculture sectors does not necessarily increase productivity. Africa's low level of human capital is particularly problematic, given the pressing need to move up the value chain from the natural resource sector into a more advanced industrial sector.

13.3 Improving Policies and Institutions in the Agricultural Sector

Adequate institution and support to agriculture have not been provided in sub-Saharan Africa. Generally, the lack of good governance and low public investment in agriculture have reduced incentives to private sector participation in agriculture. sub-Saharan African governments have spent on average less than 1% of national budgets on agriculture even if they pledged to spend 10% in 2003 under the terms of the New Partnership for Africa's Development (NEPAD)-CAADP (see introductory chapter). Building institutions to support the institutional development of rural Africa is of utmost importance. This section reviews more particularly the need for land reforms, gender-sensitive policies and farmers' organizations in agriculture.

Even though land reforms should be country-specific, Moyo et al. (2015) recommends that they clearly define property rights, ensure the security of land tenure and enable land to be used as collateral. For Kanu et al. (2014), the promotion of more equitable land access and rights requires both land registration and legal recognition of customary rights and administrative issues. If they cause land expropriation for smallholder farmers, pastoralists, indigenous communities and other vulnerable groups, land reforms will contribute to food insecurity and increase in poverty and inequality. For instance, land rights and administration have attracted attention in the context of biofuel production and land grabbing by large corporates. In Ghana, Lauretta S. Kemeze, Akwasi Mensah-Bonsu, Irene S. Egyir, D. P. K. Amegashie and Jean Hugues Nlom (Chap. 6) affirm that proper regulation would have avoided the massive conversion of fertile land to biofuel crops, at the expense of food crops. Jatropha cultivation (biofuel sector) could have been promoted on marginal lands so as to not compromise food security. Also, the lack of market for Jatropha seeds highlighted the need to properly regulate the sector in order to protect rural people.

Farmer support organizations are essential in acquiring, applying and continuously disseminating knowledge and skills to farmers (Kanu et al. 2014). Such organizations can help in mitigating risks and increasing investment. Boris Odilon Kounagbè Lokonon (Chap. 8) discusses the extent to which land tenure affects vulnerability to climate shocks, and recommends farmers' labor sharing groups and farmers' organizations. They will lessen vulnerability and raise awareness on relevant technology and good environmental management practices to increase resilience.

Besides, although women constitute most of Africa's labor force in agriculture, rules governing ownership and transfer of land rights are not favorable to women (NEPAD 2013). They face important inequalities in accessing and controlling over land, property and resources. For Kanu et al. (2014), women empowerment is needed in various forms such as policy dialogue, legal reforms, public campaigns, project development, civil society involvement and support to women groups and organizations. Land reforms should also benefit women. Adedoyin Mistura Rufai, Kabir Kayode Salman and Mutiat Bukola Salawu (Chap. 4) plead for the implementation of gender-sensitive policy. According to O. E. Ayinde, T. Abdoulaye, G. A. Olaoye and A. O. Oloyede (Chap. 5), women should be involved in the development and testing of agricultural innovation. If women farmers' preferences are incorporated in the development of agricultural technology, this will ensure food security and increased productivity.

13.4 Innovative Financing for the Transformation of African Agriculture

In sub-Saharan Africa, insufficient cash income and the difficult access to the financial and insurance sectors hampers farmers' ability to develop and to invest. Given the inadequate access to finance, in particular risk capital, subsistence farmers are not able to adopt new varieties and methodologies, hereby constraining agricultural productivity. For Moyo et al. (2015), Africa's general low financial inclusion is even worse in agriculture due to the specific production cycle.[2] The challenges of providing acceptable collateral for agricultural lending and adapting loan repayment schedules to crop cycles are huge. As a result, another means for boosting agriculture is to provide actors with the adequate financial resources.

Carren Pindiriri (Chap. 3) recommends the improvement of farmers' access to credit as it has a significant effect on farmers' decision to adopt modern technology. Financial inclusion through the establishment of rural financial institutions can significantly enhance modernization of agriculture. Regarding insurance, Francis H. Kemeze (Chap. 9) discusses the effectiveness of weather index insurance in protecting farmers against climate variability. The results are mitigated due to the specificity of the agriculture sector. Even though insurances cover basis risk, they may not cover the actual on-farm losses. Not only that they do not replace the crop loss but farmers have to buy staple food at increased price in a context of weather shocks. For this reason, the author recommends to complement weather index insurance with supplemental irrigation technology to protect farmers in case of long dry spell or severe drought.

13.5 Strengthening Africa's Agriculture Value Chains, Trade and Competitiveness

A major opportunity for Africa is to build regional value chains in the agro-industry as it will entail agriculture sector growth and job creation. For Kanu et al. (2014), economic potentials for small and medium enterprises are enormous throughout the agricultural value chain, given the untapped agro-industry market opportunity. Greater integration into value chains is expected to boost farmers' benefits and facilitate agri-business. It will eventually increase trade and integration into global value chains.

Namalguebzanga C. Kafando (Chap. 10) proposes to complete agri-business development with trade openness and regional integration policies as well as good governance to rapidly benefit from regional and global value chains. It also requires the redefinition of educational policies to efficiently use technology and the development of transport infrastructure. Infrastructure which increases agricultural productivity, reduces post-harvest loss and transportation costs, is particularly in the case of cross-border trade and value chain integration (Kanu et al. 2014). In the same vein, Anani N. Mensah and Abdul-Fahd Fofana (Chap. 12) plead for the improvement of international competitiveness and the support of export companies. There is a need to increase the sophistication and diversification of export products and to move away from forward integration In contrast, African countries should develop backward integration, which means that economies must import primary products from abroad to add value and produce high-tech products locally.

The creation of special economic zones and growth poles has been another means to encourage industrialization and attract foreign direct investment. Joseph Tinarwo (Chap. 10) recommends to integrate SEZ programs into national development strategies so as to ensure high-level political commitment. The author highlights the need to create joint ventures between foreign SEZ companies and local companies with the establishment of low minimum SEZ investment thresholds for local companies. Also, it is important to ensure infrastructure development within and outside the SEZ and to respond to SEZ labor requirements by aligning curricula of universities and Technical Vocational Education and Training (TVET) institutions.

13.6 Creative Infrastructure Solutions to Boost and Transform African Agriculture

Adequate and well-functioning infrastructure is essential for agriculture due to its positive impacts on the costs of delivering agriculture inputs and accessing market for selling outputs. However, sub-Saharan Africa's massive disadvantage in infrastructure (mainly roads, electricity and communications) has increased transaction costs and market risks, especially in small and landlocked countries. The insufficient number of roads has constituted a barrier to agricultural trade and to the adoption of productivity-enhancing inputs. For instance, Christelle Tchamou Meughoyi (Chap. 2) recommends the improvement of infrastructure to supply farmers with the fertilizers they need to adopt improved maize seeds.

Infrastructure is also key to reduce the reliance on rain-fed agriculture and to deal with climate variability through increased irrigation. Idrissa Ouiminga (Chap. 7) highlights the positive effects of soil and water conservation techniques as an alternative to adapting to climate change. Traditional constructed structures or dug are effective to retain water and financially affordable. Zaï (seed holes dug perpendicularly to the slope and staggered), stony ropes or half-moons help farmers to combat land degradation and desertification.

Notes

1. The Green Revolution refers to the use of pesticides, the better management techniques and the introduction of improved varieties of cereals which allowed Asian countries to double the cereal production between 1970 and 1995, whereas the total land area cultivated with cereals increased by only 4%. For further information on the Green Revolution, refer to Moyo et al. (2015).
2. The production cycle in agriculture consists of an initial high investment then a long period of no cash inflows during the growing season and finally, a large cash windfall after harvest, except in the case of natural or weather disaster.

References

AfDB, Organisation for Economic Co-operation and Development [OECD], & United Nations Development Programme [UNDP]. (2017). *African Economic Outlook 2017*. Paris: OECD Publishing. http://www.africaneconomicoutlook.org/en/.

Kanu, B. S., Salami, A. O., & Numasawa, K. (2014). *Inclusive Growth: An Imperative for African Agriculture*. Tunis: African Development Bank.

Moyo, J. M., Bah, E. M., & Verdier-Chouchane, A. (2015). Transforming Africa's Agriculture to Improve Competitiveness. In World Economic Forum, World Bank and AfDB, *Africa Competitiveness Report 2015*. Geneva: WEF.

NEPAD. (2013). *Agriculture and Africa – Transformation and Outlook*. Johannesburg: NEPAD.

Verdier-Chouchane, A., & Karagueuzian, C. (2016). Moving Towards a Green Productive Agriculture in Africa: The Role of ICTs. *Africa Economic Brief* *7*(7). Côte d'Ivoire: African Development Bank.

Index[1]

[1] Note: Page numbers followed by 'n' refer to notes.

A. Shimeles et al. (eds.), *Building a Resilient and Sustainable Agriculture in Sub-Saharan Africa*, https://doi.org/10.1007/978-3-319-76222-7

J

K

L

M

N

www.ingramcontent.com/pod-product-compliance
Lightning Source LLC
LaVergne TN
LVHW082117200926
848484LV00008B/1967